Plasmonic Nanosensors for Detection of Aqueous Toxic Metals

W0235013

Plasmonic Nanosensors for Detection of Aqueous Toxic Metals

Edited by
Dinesh Kumar

With contributions from
Rekha Sharma

CRC Press
Taylor & Francis Group
Boca Raton London New York

CRC Press is an imprint of the
Taylor & Francis Group, an **informa** business

First edition published 2022
by CRC Press
6000 Broken Sound Parkway NW, Suite 300, Boca Raton, FL 33487-2742

and by CRC Press
2 Park Square, Milton Park, Abingdon, Oxon, OX14 4RN

CRC Press is an imprint of Taylor & Francis Group, LLC

© 2022 selection and editorial matter, Dinesh Kumar; individual chapters, the contributor.

Reasonable efforts have been made to publish reliable data and information, but the author and publisher cannot assume responsibility for the validity of all materials or the consequences of their use. The authors and publishers have attempted to trace the copyright holders of all material reproduced in this publication and apologize to copyright holders if permission to publish in this form has not been obtained. If any copyright material has not been acknowledged please write and let us know so we may rectify in any future reprint.

Except as permitted under U.S. Copyright Law, no part of this book may be reprinted, reproduced, transmitted, or utilized in any form by any electronic, mechanical, or other means, now known or hereafter invented, including photocopying, microfilming, and recording, or in any information storage or retrieval system, without written permission from the publishers.

For permission to photocopy or use material electronically from this work, access www. copyright. com or contact the Copyright Clearance Center, Inc. (CCC), 222 Rosewood Drive, Danvers, MA 01923, 978-750-8400. For works that are not available on CCC please contact mpkbookspermissions@ tandf.co.uk

Trademark notice: Product or corporate names may be trademarks or registered trademarks and are used only for identification and explanation without intent to infringe.

ISBN: 9780367651923 (hbk)
ISBN: 9780367651930 (pbk)
ISBN: 9781003128281 (ebk)

DOI: 10.1201/9781003128281

Typeset in Times
by codeMantra

Contents

Acknowledgments

First and foremost, praise and thanks to the Almighty for blessing me to complete the book throughout my work. Writing a book is more challenging than I thought and more rewarding than I could have ever imagined. None of this would have been possible without my research supervisor Prof. Dinesh Kumar, School of Chemical Sciences, Central University of Gujarat, Gandhinagar, India. I want to express my deep and sincere gratitude to him for allowing me to do research and providing invaluable guidance throughout this work. His dynamism, vision, sincerity, and motivation have deeply inspired me. He has taught me the methodology to carry out the work and present this book's writing as clearly as possible. It was a great privilege and honor to work and study under his guidance. I am incredibly grateful for what he has offered me. I would also like to thank him for his friendship, empathy, and great sense of humor.

Last, but not the least, I am extremely grateful to my parents for their love, prayers, caring, and sacrifices for educating and preparing me for my future.

Dr. Rekha Sharma, PhD
Assistant Professor in the Department of Chemistry, Banasthali Vidyapith

Authors

Dr. Dinesh Kumar is a Professor in the School of Chemical Sciences at the Central University of Gujarat, Gandhinagar, India. Prof. Kumar earned his Master's and PhD in Chemistry from the Department of Chemistry, University of Rajasthan, Jaipur. Prof. Kumar has received many national and international awards and fellowships. His research areas of interest include developing capped MNPs, core–shell NPs, and biopolymers incorporated metal oxide-based nanoadsorbents and nanosensors to remove and sense health-hazardous inorganic toxicants and heavy metal ions from aqueous media. For water purification, Prof. Kumar developed hybrid nanomaterials from different biopolymers such as pectin, chitin, cellulose, and chitosan. His research interests also focus on the synthesis of supramolecular metal complexes; metal chelates and their biological effectiveness. He has authored or co-authored more than 115 publications in journals of international repute, two books, more than seven dozen book chapters, and ~100 presentations/talks at national/international conferences.

Dr. Rekha Sharma earned her BSc from the University of Rajasthan, Jaipur, in 2007. In 2012, she completed her MSc in Chemistry from Banasthali, Vidyapith. She earned her PhD in 2019 from the same university, under the supervision of Prof. Dinesh Kumar. Presently, she is an Assistant Professor in the Department of Chemistry, Banasthali Vidyapith, and has entered into a specialized research career focused on developing water purification technology. With three years of teaching experience, she has published seven articles in journals of international repute and more than 25 book chapters in the field of nanotechnology. She has presented her work at more than 15 national and international conferences. Her research areas of interest include developing water purification technology by the fabrication of nanosensors and nanoadsorbents for water and wastewater treatment.

Abbreviations

0D	Zero-dimensional
1D	One-dimensional
2D	Two-dimensional
3D	Three-dimensional
AA	Ascorbic acid
AAS	Atomic absorption spectrometry
AES	Atomic emission spectrometry
AFM	Atomic force microscopy
AFS	Atomic fluorescence spectrometry
AgNPs	Silver nanoparticles
ASV	Anodic stripping voltammetry
ATR-IR	Attenuated total reflection infrared
AuNPs	Gold nanoparticles
BET	Brunauer–Emmett–Teller
CDs	Carbon dots
CE	Capillary electrophoresis
CEA	Carcinoembryonic antigen
CL	Chemiluminescence
CNDs	Carbon nitride dots
CNs	Carbon nanodots
CNTs	Carbon nanotubes
CPEs	Conjugated polyelectrolytes
CQDs	Carbon quantum dots
CSDTC	Chitosan dithiocarbamate
CTAB	Hexadecyltrimethylammonium bromide
CTP	Cytidine triphosphate
CURN	Curcumin nanoparticles
CV	Cyclic voltammetry
CVD	Chemical vapor deposition
Cz-TPM	Carbazolic porous framework
DFT	Density functional theory
DLS	Dynamic light scattering
DMG	Dimethylglyoxime
DO	Dissolved oxygen
DOPA	3,4-dihydroxyphenylalanine
DPASV	Differential pulse anodic stripping voltammetry
dsDNA	Double-stranded DNA
DTC	Dithiocarbamate
DTPDI	Dithioacetal-functionalized perylenediimide
DTT	Dithiothreitol
EDS	Emission-dispersive X-ray spectroscopy

EIS	Electrochemical impedance spectroscopy
ELISA	Enzyme-linked immunosorbent assay
EM	Electromagnetic
EMR	Electromagnetic radiation
EPA	Environmental Protection Agency
ESNFs	Electrospun nanofibers
FBG	Fiber Bragg grating
FESEM	Field emission scanning electron microscopy
FET	Field-effect transistor
FRET	Fluorescence resonance energy transfer
FTIR	Fourier transform infrared spectroscopy
FWHM	Full width at half maximum
GCE	Glassy carbon electrode
GNMACPE	Anthocyanin capped carbon paste electrode
GNRs	Gold nanorods
GO	Graphene oxide
GQDs	Graphene quantum dots
GSH	Glutathione
HA-AgNPs	Humic acid-functionalized silver nanoparticles
Hcy	Homocysteine
HMs	Heavy metals
ICP-MS	Inductively coupled plasma mass spectroscopy
ICP–OES	Inductively coupled plasma–optical emission spectroscopy
ITO	Indium tin oxide
LBL	Layer-by-layer
LDA	Linear discrimination analysis
LIBS	Laser-induced breakdown spectroscopy
LOB	Limit of blank
LOD	Limit of detection
LOQ	Limit of quantification
LSPR	Localized surface plasmon resonance
MALDI-MS	Matrix-Assisted Laser Desorption/Ionization mass spectrometry
MDP	Methylenediphosphonic acid
MNPs	Metal nanoparticles
MP	Microprobes
MPA	Mercaptopropionic acid
MS	Mass spectrometry
MSA	2-mercaptosuccinic acid
MUA	11-mercaptoundecanoic acid
MWCNTs	Multi-walled carbon nanotubes
NaCysC	N-cholyl-L-cysteine
NCF	No-core fiber
NEAs	Nanoelectrode arrays
NFM	Nanofibrous membrane

NIR	Near-infrared
NPs	Nanoparticles
NPSi	Nanoporous silicon
OPD	O-phenylenediamine
PAA	Porous anodized aluminum oxide
PAN	Polyacrylonitrile
PANI	Polyaniline
PDs	Polymer dots
PEC	Photoelectrochemical
PEF	Plasmon-enhanced fluorescence
PEG-PVP	Polyethylene glycol poly vinylpropyl
PESs	Paper-based electrochemical sensors
PET	Polyethylene terephthalate
PL	Photoluminescence
PLQY	Photoluminescence quantum yield
PMMA	Polymethylmethacrylate
POC	Point-of-care
POFE	Palm oil fronds extracts
PPE	Poly(p-pheynylene ethynylene)
PSPR	Propagating surface plasmon resonance
PSS	Polystyrene sulfonate
QY	Quantum yield
R6G	Rhodamine 6G
Rac	Ractopamine
RET	Resonance energy transfer
RF	Removal efficiency
rGO	Reduced graphene oxide
rGO/PEI/Pd	Reduced graphene oxide/polyethylenimine/PdNPs
RhB	Rhodamine B
ROS	Reactive oxygen species
RSD	Relative Standard Deviation
RTP	Room temperature phosphorescence
SBR	Smartphone-based portable reader
SE	Solid-phase extraction
SEIAS	Surface-enhanced infrared absorption spectroscopy
SERS	Surface-enhanced Raman scattering
SiO$_2$@Au	Silica-gold core-shell
SMF	Single-mode fibers
SP	Surface plasmon
SPE	Screen-printed electrodes
SPPs	Surface plasmon polaritons
SPR	Surface plasmon resonance
ssDNA	Single-stranded DNA
SWASV	Square wave anodic stripping voltammetry
SWCNTs	Single-walled carbon nanotubes

SWSV	Square wave stripping voltammetry
SWV	Square wave voltammograms
TCA	Thiacalixarene
TEM	Transmission electron microscopy
TGA	Thermo-gravimetric Analysis
TMPM	3-(Trimethoxysilyl) propyl methacrylate
TMPyP	5,10,15,20-tetrakis(1-methyl-4- pyridinio) porphyrintetra(p-toluenesulfonate)
UV-Vis	Ultraviolet-visible
WHO	World Health Organization
XFS	X-ray fluorescence spectroscopy
XPS	X-ray photoelectron spectroscopy
XRD	X-ray powder diffraction

1 Plasmonic Nanosensors
An Introduction

1.1 INTRODUCTION

In metals, plasmons are the collective oscillation of conduction band electrons, which can also be defined as a quantized form of plasma oscillation. The collective oscillation of the conduction band electrons is also called surface plasmon resonance (SPR). The electrons resonate with the oscillating electric field of incoming electromagnetic (EM) radiations, generating energetic plasmonic electrons through non-radiative excitation [1]. The position of SPR, together with its intensity strappingly, depends on the neighboring atmosphere's dielectric properties and the size, shape, and composition of the nanostructures [2,3]. The optical sensors can be developed using plasmonic metallic nanostructures using this type of responsive variable. Hence, in recognition of analyte in homeland security, biomedical diagnosis, food safety, and environmental monitoring, plasmon-enhanced optical sensors show augmenting usage [4,5]. The two forms of SPR are as follows: localized surface plasmon resonance (LSPR) and propagating surface plasmon polaritons (SPPs). In the metallic nanostructure, LSPR arises from collective and non-propagating oscillations of surface electrons when the dimensions are less than the wavelength of incident light. These nanostructures have been used to develop SPR-based colorimetric nanosensors to detect a toxic analyte of water at a trace level. Supplying the source for colorimetric plasmonic sensors, the LSPR strappingly hinges on the neighboring medium's refractive index. LSPR around the nanostructure also concentrates the incident EM field. The optical processes, such as infrared absorption, Raman scattering, plasmon-enhanced fluorescence (PEF), surface-enhanced infrared absorption spectroscopy (SEIAS), and surface-enhanced Raman scattering (SERS), have been widely used for developing plasmonic nanosensors. For resonance excitation, SPPs necessitate matching momentum, for example, through periodicity in a nanostructure, and could not be excited through free-space radiation.

SPPs are transduced to the sensor's signal and are modulated through the refractive index of the surroundings. In modulating radiation in SERS and PEF, SPP can likewise perform an important role. Through alteration at a distance farther from the nanostructure's surface, the fleeting EM field of SPP permitting the SPP to be modulated, which declines with an elongated length (mostly 200 nm) compared to LSPR. Several plasmonic metallic nanostructures as signal transducers and amplifiers have already been established for sensitive optical sensing through SPP and LSPR. This chapter mainly focuses on the principles and concepts of plasmonics; this chapter also discusses three major sorts of optical

DOI: 10.1201/9781003128281-1

sensors based on plasmonic nanostructures, namely SERS, PEF, and plasmonic sensors. Similarly, this chapter emphasizes the principles to summarize the fundamental physics for designing diverse plasmon-enhanced optical sensors and optimization methods that augment signal amplification and transduction.

1.2 SURFACE PLASMON RESONANCE: FUNDAMENTAL PRINCIPLES

In any metal, a time-dependent force influences the free conduction electrons, which occur contradictory to alter the EM field of the incident light (Figure 1.1a). Owing to the dampening caused by Ohmic losses and the electron's charge, the electrons' subsequent motion is 180° out of phase and oscillatory [6]. Like all oscillators, the plasma frequency is given as equation 1.1, which is also known as the characteristic frequency (ωp) of conduction electrons.

$$\omega p = \sqrt{\frac{ne^2}{m_{\text{eff}\varepsilon_o}}} \tag{1.1}$$

In response to the incident field, corresponding to the movement of electrons, the plasma frequency hinge onto the effective mass (m_{eff}) and the density of electrons (n). Moreover, the permittivity of free space is represented as ε_o and the charge of an electron is represented as e [6]. Usually, there is not a single resonant frequency due to unlike mass on a spring. Therefore, in contrast to a restoring force, the free conduction electrons do not oscillate in the bulk of the metal. As an alternative, based on mass being dragged in the electrons or a viscous fluid, the motion will vary that can respond quickly enough to the driving force of the incident field. If the light has a frequency above the plasma frequency, the light will merely be absorbed or transmitted in interband transitions, and at this frequency, the electrons will not oscillate [6]. The electrons will oscillate 180° out of phase if the light has a frequency smaller than the UV range, which causes a strong reflection of the incident light [6]. The metals attain their characteristic color by interband transitions and the combination of plasma frequency. By the use of equation 1.2 of the real part of the dielectric constant ($\varepsilon'_{\text{metal}}$), this behavior can be described mathematically [7].

$$\frac{\varepsilon'_{\text{metal}}}{\varepsilon_o} = 1 - \left(\frac{\omega p}{\omega}\right) \tag{1.2}$$

By the positive value of the real part of the dielectric constant, the light is transmitted, which occurs when the plasma frequency is lesser than the frequency of light [7]. By the negative value of the real part of the dielectric constant, most of the light is reflected, which occurs when the plasma frequency is higher than the frequency of light (Figure 1.1d and e) [7]. The dielectric constant decides whether or not the metal electrons can oscillate at the given frequency of light. Figure 1.1 demonstrates that the Ohmic losses occur because of the positive

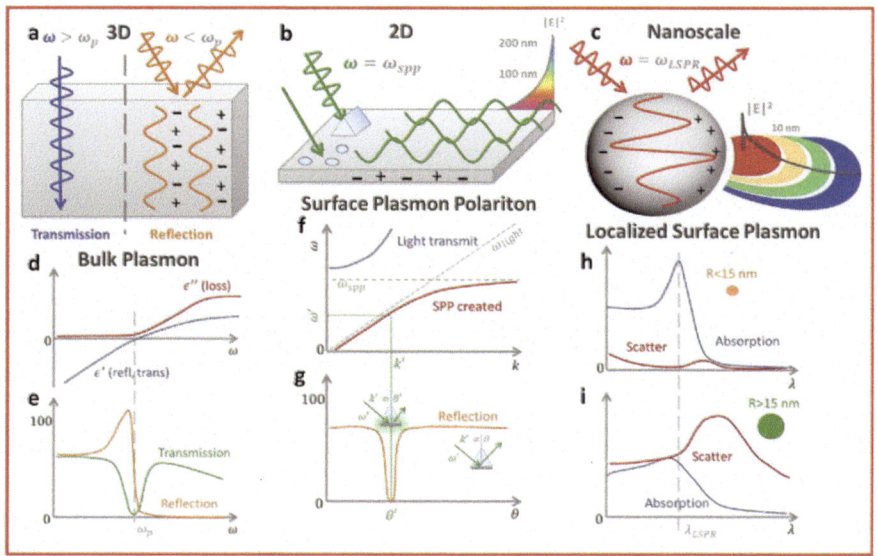

FIGURE 1.1 Diagrammatic representation of LSPR, surface, and volume. (Reprinted with permission from Ref. [7]. Copyright 2015 The Royal Society of Chemistry.)

unreal part of the dielectric constant. Corresponding to proliferating charge waves, the SPPs oscillate at the surface if the bulk metal is now shrunk to a thin film (Figure 1.1b) [8,9].

The itinerant surface charge waves through volume oscillations and the transformation of the bulk are known as polariton. In the incident field, the frequencies of the metal electrons on which they can oscillate depend on the surrounding medium (ε_{diel}) and the interface among the metal (ε_{metal}). This confines the continuous spectrum of equation 1.2 for a given interface to a fixed frequency and wave vector for all incident angles. The dispersion curve for the SPP is demonstrated by equation 1.3, which gives the resonance condition to excite the SPP [8,9].

$$K_{ssp} = \frac{\omega}{C} \sqrt{\frac{\varepsilon_{metal}\varepsilon_{diel}}{\varepsilon_{metal}+\varepsilon_{diel}}} \tag{1.3}$$

The light wave vector for an interface is essential to stimulate an SPP displayed by the dispersion curve (Figure 1.1f). The massless photon is constantly lesser than the wave momentum or vector of the oscillating charge wave [8,9]. Consequently, SPP cannot be directly excited to supply the extra momentum by the incident light. In contrast, it can be excited utilizing a grating or through a prism in the Kretschmann geometry [8]. To excite the SPP, the dispersion curve gives the angle in equation 1.3, for which the prism or the grating can supply the required momentum. Corresponding to a dip in the transmission or reflection spectrum, the light will be absorbed at this angle (Figure 1.1g). Through canceling out the denominator in equation 1.3, the actual part of the dielectric constant is negative when the metal electrons oscillate, which corresponds to a resonance condition at [8,9].

$$\omega_{ssp} = \frac{\omega_p}{\sqrt{1 + \varepsilon_{diel}}} \qquad (1.4)$$

For a sensor on the dielectric constant through the dependency of the SPP frequency, the bulk plasma oscillations at the interface transformed into a helpful transducer. The local EM field encompasses 100–200 nm hooked on the dielectric ensuing through the charge oscillations of the SPP [10]. The dielectric constant will vary when the local circumstances vary within this distance, and the resultant SPP frequency will shift in the air from that distance. At the interface, this can be implicitly abstractly reducing repulsion among adjacent electrons and as the dielectric screening of the charge, red shifting the oscillation frequency and efficiently dropping the energy required to drive oscillations. For SPP-based sensors, in height, angular specificity of the SPP and the narrow absorption line shape consents outstanding figures of merit to be attained and signal-to-noise ratio [8]. The sensitivity comes from the complex geometries required for recognition at the trade-off of experimental easiness [8]. The limitations of SPP could be overwhelmed through forming a zero-dimensional (0D) nanoparticle from a two-dimensional (2D) metal film. From the positive ionic core upbringing, persuading a sturdy restoring force and a consistent displacement of the electron density, if the incident electric field is lesser than the wavelength of light, it will be constant across the nanoparticle (Figure 1.1c) [11]. Like a simple harmonic oscillator, the restoring force in the metal electrons corresponds to a distinctive oscillation frequency known as LSPR [11]. When LSPR can be thrilled directly through the incident field, it supplies the additional momentum because of the nanoparticle's geometry [11]. In the LSPR peak position, the local environment-induced change deprived of the requirement for added prisms or gratings can be sensed utilizing a simple ultraviolet-visible (UV-Vis) spectrometer (Figure 1.1h). As the extinction (scattering + absorption) cross-section is articulated using a simple harmonic oscillator model or the Mie theory, the precise circumstances for LSPR could be solved for a nanosphere [12–14].

$$\sigma_{ext} = 9\left(\frac{\omega}{C}\right)\left(\varepsilon_{diel}\right)^{\frac{3}{2}} V \frac{\varepsilon''_{metal}}{\left(\varepsilon'_{metal} + 2\varepsilon_{diel}\right) + \left(\varepsilon''_{metal}\right)^2} \qquad (1.5)$$

Equation 1.5 demonstrates that corresponding to a strong resonance condition, and the denominator will vanish when the factual part of the dielectric function is negative. The electrons in the metal oscillate can shift with variation in the local dielectric circumstances, as given in equation 1.6.

$$\omega_{LSPR} = \frac{\omega_p}{\sqrt{\left(1 + 2\varepsilon_{diel}\right)}} \qquad (1.6)$$

At resonance, the scattering and absorption cross-section occurs by the articulate oscillations of the electrons. Numerous orders of magnitude are given by $V=4/3\pi R^3$, which are larger than the nanoparticle's physical size.

Figure 1.1a shows the plasma frequency of metal, which is the frequency below which the conduction electrons can oscillate in the incident field. Because of a negative real part of the dielectric constant (d), these oscillations occur together with the increased reflection from the metal (e). SPPs (b) are the electron oscillations corresponding to a propagating charge wave on a 2D surface. These oscillations propagate along with the interface and exponentially decrease away from the interface. These are coupled with amplitude and an EM field. The SPP exists as a field that declines fleetingly from the surface and can only be excited at certain wave vectors (f). The matching condition with the momentum existing at certain incident angles corresponds to the SPP resonance (g). LSPR exists when the incident wavelength is larger than the metal nanoparticle (c), corresponding to in-phase electronic oscillation. The collective oscillations and an amplified local EM field correspond to a scattering cross-section and higher absorption. The absorption cross-section is large, and for small particles less than 15 nm, the absorption dominates (h). For larger nanoparticles (NPs) of more than 15 nm, the scattering cross-section dominates. The EM field is taken as polarized as shown in Figure 1.2 in the plane of incidence.

Several vital changes have been considered, which occur among LSPR and SPP when designing a sensor. Before the interfacial dielectric constant, the factor of 2 depends on the nanoparticle's geometry. As the metal used is altered with the shape and the surroundings, the LSPR peak position will fluctuate (Figure 1.2) [15–17]. As the size of the nanoparticle upsurges, the plasmon further red-shifted at opposite surfaces and concomitant the reduction of repulsion among electrons [18–20]. In numerous orders of magnitude, ensuing an intense local EM field could be sturdier than the incident field strength brought about by the limited electron oscillations in LSPR.

Further, through improving the sensitivity and increasing the local field intensity, the field will be concentrated to changes in the local environment analogous to a lightning rod with sharp edges in NPs [21,22]. In contrast to SPP, the

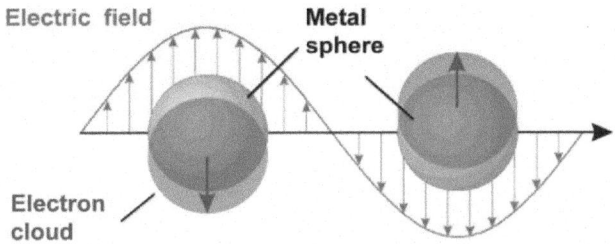

FIGURE 1.2 Diagrammatic representation of LSPR under the electric field where coherent electrons oscillate around the nanoparticle surface. (Reprinted with permission from Ref. [10]. Copyright 2017 American Chemical Society.)

local refractive index of the neighboring circumstances decays in 100–200 nm. Consequently, from the surface of the metal, the EM field in LSPR is more sensitive to change in the distance and decays in 10–30 nm [8,9]. Into the far-field-like scattering, the plasmon can also re-radiate its energy. If scattering or absorption dominates with particle size determination, subsequently, the LSPR can be thrilled through incident light (Figure 1.1h and i) [16]. The energy of the LSPR can be rapidly adapted into heat through the electron–electron scattering, translating a strong absorption for less than 15 nm metal NPs (MNPs) [23,24]. The plasmon's energy in larger particles leads to a sturdy scattering cross-section when the electron–electron surface scattering is condensed [23,24]. The SPP has a higher lifetime than the LSPR because of the electron–electron scattering and radiative dampening [25]. SPP has a narrow absorption peak than LSPR since the spectral width is contrariwise associated with the lifetime, reducing the competence of these sensors. By utilizing planar arrays of MNPs and optimizing their geometry, the absorption linewidth could be enhanced, declining this drawback. Finally, due to the same local fields of LSPR and SPP, the coupling can occur among LSPR/LSPR, SPP/LSPR, and SPP/SPP when the supporting metal structures are brought within the local field deterioration length [13]. Because of hybridization among the modes, the coupling can be principal to a shift in the spectral position and an improved local field [26]. For instance, the increase in the local field of two spheres when they are aggregated goes from 10 to 10^4 [13]. Consenting lower detection levels than film-based designs or single particles, a shift in the LSPR or SPP frequency can be attained upon adding an analyte through accumulating numerous plasmonic structures.

1.3 PLASMONIC NANOPARTICLE-ASSOCIATED COLORIMETRIC SENSORS

Under the stimulation of light, the method driven by the oscillation of free electrons on a metal surface is called plasmonics [27]. In contrast to the attraction force to their positive nuclei, the SPR occurs when resonance can be attained through the incident light having the same frequency in the form of the oscillating surface electron (Figure 1.2) [28]. Characteristically, two SPR modes have been discovered in plasmon, and these modes are highly confined to 2D nanostructured arrays and the shape of the colloidal nanostructures. The first one represents propagating surface plasmon resonance (PSPR) at the flat dielectric-metal interface, and another one is LSPR [28–31]. The surface EM waves can be propagated at a metal surface through PSPR. The LSPR produces exceedingly improved and confined EM fields on the boundary of nanoparticle surfaces in plasmonic nanostructures particularly developed from silver and gold. This method is done in cooperation with prominent absorption maxima in the frequency range of near-infrared (NIR) and visible [32]. This indicates that the oscillations can stimulate by the interaction among chemical entities and NPs and the NPs' shape, size, dielectric environments, and surface coating. These events can be altered by scattering or extinction/absorption into a color variation of the plasmonic nanoparticle solutions and LSPR spectral change [29]. The position of LSPR of plasmonic NPs is shifted through the introduction of the targets. These astonishing plasmonic methods increase to a fast-emerging field in colorimetric nanosensors. The solution's color change is done by transforming morphology/size or forming a cross-linking aggregate/assembly of plasmonic NPs. These ultrasensitive assays will be abridged into two types according to the diverse signal producing mechanism:

  i. Colorimetric assay based on inter nanoparticle distance and
  ii. Colorimetric assay based on NPs' size/morphology.

1.3.1 COLORIMETRIC ASSAY BASED ON INTER NANOPARTICLE DISTANCE WITH PLASMONIC NANOPARTICLES

Accompanied by the visible color transition and LSPR shift of nanoparticle solution, the aggregation of plasmon NPs will induce surface plasmon coupling with suitable sizes (diameter > 3.5 nm) among the neighboring particles [33]. For instance, the color transition from red to the blue of the gold NPs (AuNPs) solution leads to the redispersion (or aggregation) actions [34]. So, for sensing any targets through the colorimetric assays, the plasmonic nanomaterials with the interparticle distance-dependent optical property offer a practical basis, such as small molecules, metal ions, living cells, or macromolecules [32,35–40]. Numerous covalent/non-covalent forces, comprising hydrophobic forces, electrostatic interactions, specific biological bonding, and hydrogen bonding, are outstandingly moderated between plasmonic NPs [41–45]. These forces are pretentious and interrelated through numerous exterior factors, such as buffer solution, pH, and temperature, throughout the driving plasmonic nanoparticle interaction. Therefore, it will challenge each inter nanoparticle interaction to explain the basic principles [41]. In colorimetric sensing applications, here we will only highlight the technology of using the driving forces to show the important role of surface coating/modification of plasmonic NPs and control the inter-distance of the NPs.

1.3.1.1 Electrostatic Interactions

To develop the assembled plasmonic nanostructure, the frequently used method is an electrostatic interaction. The electrostatic interactions can facilitate nanoparticle assembly when surfaces of plasmonic NPs, for example, AgNPs and AuNPs, are electrically charged. The external stimuli are comprising the addition of precise biomolecular species, for example, DNA, proteins. Along with many neighboring environments, for example, ionic strength, pH, light, electrical field, mechanical stress, the temperature can control redispersion or aggregation of NPs and influence the nanoparticle surface under purely electrostatic interactions, resulting in an LSPR change. Various research studies have reported on electrostatic interactions based on these stimuli-responsive abilities that intricate colorimetric assay based on plasmonic interactions [32,35,38,39]. However, because of the inherent weak force of electrostatic interactions, there is a drawback: for direct quantification, the detection sensitivity is generally inadequate.

In addition, to increase the interaction forces and progress the sensitivity throughout nanoparticle aggregation actions, many other methodologies have been proposed except for using multivalent electrostatic interactions. They predominantly combine highly efficient physicochemical, polymerization, and other enzymatical catalysis reactions [46–60].

1.3.1.2 Covalent Bonding

Multilayer structure and covalent bonding can empower the development of steady definite assemblages. Furthermore, this intermolecular interaction permits numerous other functional groups to integrate onto nanoparticle surfaces through excess reactive groups. The covalent bonding can offer advanced binding force between the building blocks, although its deficiencies suppleness in the self-assembly method compared to other forces. Covalent bonding shorter the detection time and enhances the sensitivity in the colorimetric assay, and thus enhances assembly efficiency [47,61]. The use of functionalized plasmonic NPs can be eased by the adaptable surface chemistry of covalent bonding [35,54].

1.3.2 PLASMONIC NANOPARTICLE COLORIMETRIC ASSAY BASED ON NANOPARTICLES' SIZE/MORPHOLOGY

For the colorimetric sensing in the target-mediated molecular events, the inter nanoparticle distance-dependent mechanism, for example, target-triggered interparticle aggregation or crosslinking. The specific alteration of the size and shape/morphology of the target-mediated molecular events plasmonic NPs have also been applied. Using some key parameters for growth control, many reports specifically yield differentiable features in the optical properties; a tiny change will cause different NPs and, mainly, ensuing in the LSPR. The capping agents, reducing agents, and precursor are the synthetic parameters, which can alter the nanoparticle growth. So, through tuning the growth of plasmonic NPs, a variation of solution color and LSPR can be easily identified together with the kinetics of etching. Therefore, some target-induced detection actions are used to control the size-dependent properties. In view of the progressive method, the morphology and size of NPs will be categorized, and overview of the related colorimetric nanosensors:

  i. growth-dependent target-induced plasmonic nanosensors, and
  ii. etching-dependent colorimetric assays.

1.3.3 TARGET-INDUCED PLASMONIC NANOPARTICLES ETCHING FOR COLORIMETRIC ASSAYS

In the plasmon-based colorimetric sensing systems, controllable chemical etching has been used to control the uses of the nanostructures in optical, catalysis, and electronic sensing as one of the most vital methods. Generally, corresponding to the alteration of the NPs' morphology/composition, which can be sensed in the etchant's presence, it comprises the oxidation of plasmonic NPs. For instance, to yield silver nanodisks in the existence of halides or H_2O_2, the etching of silver triangular nanoplates can reshape. By the change in shape to the rounded corners of spherical NPs from the sharp edges of the triangles, depending on the concentration of the analyte, the color of the solution altered from blue/purple to yellow or red-colored solution.

1.4 CONCLUSIONS AND FUTURE PERSPECTIVES

Plasmonics can be used to augment SERS and arbitrate fluorescence or used directly for signal transduction. Sensors are well recognized based on plasmon-transduction. Especially, colorimetric sensors can be effortlessly built, and these are simple. Though meddling analogs to SERS and fluorescence, this type of sensor has comparatively more vulnerable and is of low sensitivity. Because of their copious lucrative accessibility, affordable price, and high sensitivity, fluorescence sensors have been extensively utilized in many fields based on visible light fluorophores. However, fluorophores have a very low quantum yield (QY) near-infrared, restraining their expediency and sensitivity. To enhance the fluorescence by incorporating a plasmonic nanostructure with the near-infrared, this can be overcome. SERS sensors are comparatively novel in contrast to plasmonic and fluorescence sensors. To consenting detection of real-world samples, in vivo imaging, and upholding outstanding anti-interference belongings, these have an excellent competence than other optical techniques at larger penetration depths to perform imaging and chemical analysis. Colorimetric plasmonic sensing and fluorescence systems could be effortlessly integrated hooked on a compact instrument.

In point-of-care (POC) devices, tapping a limitation on the application of SERS, these instruments were historically expensive and large. To solve this problem, extensive progress has been made in the last decade. Palm-sized Raman readers or commercial benchtop are accessible nowadays. To additionally enhance the accessibility of compact SERS sensing systems, the progress of plasmon-enhanced SERS technology poises. Nowadays, owing to conformability, flexibility, efficient transport, and efficient uptake of analyte because of high specific surface area and hierarchical vasculature, paper- and fiber-based SERS substrates have attracted the attention of SERS research [62–65]. These innovative SERS substrates are likewise effortlessly integrated hooked on microfluidics, conventional chromatography, and other biological assays, consequently grasping great promise in future SERS applications. In addition, the capability of multiplexed detection in live animals is one of the most promising features of future SERS [66–69].

Consequently, SERS as the plasmonic-based design improves great promise for in vivo biological imaging and chemical sensing. Based on SERS and fluorescence devices in the future, plasmonics will remain to be developed to design new detection schemes and enhance the sensing signal. So far, on a practical foundation, various plasmon intricate sensors have been developed. To confirm the sustained progress of virtuous plasmon-mediated sensors, a "Device-by-Design" method is required. But deprived of first establishing an in-depth understanding of the basic theory, this is impossible. The mechanisms of plasmon-enhanced SERS and plasmonic transduction have previously been well studied. Though inconsiderate the basic mechanism of plasmon-mediated fluorescence, substantial advancement has been completed. To elucidate the misperception of enrichment versus quenching, further efficient studies are essential to accomplish. When the basic theory is well understood, the confines of plasmonic sensors can only be pushed. For designing plasmon-enhanced sensors by providing flexibility and prodigious opportunity, both SPP and LSPR can be tailored to the use of nanostructured architecture and materials. The development of new methods and the effective design for the fabrication of plasmonic materials/architectures is the success of future plasmon-enhanced sensors.

REFERENCES

1. Jones, Matthew R., Kyle D. Osberg, Robert J. Macfarlane, Mark R. Langille, and Chad A. Mirkin. "Templated techniques for the synthesis and assembly of plasmonic nanostructures." *Chemical Reviews* 111, no. 6 (2011): 3736–3827. doi:10.1021/cr1004452.
2. Ringe, Emilie, Mark R. Langille, Kwonnam Sohn, Jian Zhang, Jiaxing Huang, Chad A. Mirkin, Richard P. Van Duyne, and Laurence D. Marks. "Plasmon length: A universal parameter to describe size effects in gold nanoparticles." *The Journal of Physical Chemistry Letters* 3, no. 11 (2012): 1479–1483. doi:10.1021/jz300426p.
3. Ghosh, Sujit Kumar, and Tarasankar Pal. "Interparticle coupling effect on the surface plasmon resonance of gold nanoparticles: From theory to applications." *Chemical Reviews* 107, no. 11 (2007): 4797–4862. doi:10.1021/cr0680282.
4. Android, Iuliana E., Megan E. Warner, and Robert M. Corn. "Fabrication of silica-coated gold nanorods functionalized with DNA for enhanced surface plasmon resonance imaging biosensing applications." *Langmuir* 25, no. 19 (2009): 11282–11284. doi:10.1021/la902675s.
5. Lieberman, Itai, Gabriel Shemer, Tzipi Fried, Edward M. Kosower, and Gil Markovich. "Plasmon-resonance-enhanced absorption and circular dichroism." *Angewandte Chemie* 120, no. 26 (2008): 4933–4935. doi:10.1002/ange.200800231.
6. Li, Ming, Scott K. Cushing, and Nianqiang Wu. "Plasmon-enhanced optical sensors: A review." *Analyst* 140, no. 2 (2015): 386–406. doi:10.1039/C4AN01079E.
7. Oates, T. W. H., Herbert Wormeester, and Hans Arwin. "Characterization of plasmonic effects in thin films and metamaterials using spectroscopic ellipsometry." *Progress in Surface Science* 86, no. 11–12 (2011): 328–376. doi:10.1016/j.progsurf.2011.08.004.
8. Lakowicz, Joseph R. "Plasmonics in biology and plasmon-controlled fluorescence." *Plasmonics* 1, no. 1 (2006): 5–33. doi:10.1007/s11468-005-9002-3.
9. Dostálek, Jakub, and Wolfgang Knoll. "Biosensors based on surface plasmon-enhanced fluorescence spectroscopy." *Biointerphases* 3, no. 3 (2008): FD12–FD22. doi:10.1116/1.2994688.
10. Tang, Longhua, and Jinghong Li. "Plasmon-based colorimetric nanosensors for ultrasensitive molecular diagnostics." *ACS Sensors* 2, no. 7 (2017): 857–875. doi:10.1021/acssensors.7b00282.
11. Mayer, Kathryn M., and Jason H. Hafner. "Localized surface plasmon resonance sensors." *Chemical Reviews* 111, no. 6 (2011): 3828–3857. doi:10.1021/cr100313v.

12. Li, Yang, Ke Zhao, Heidar Sobhani, Kui Bao, and Peter Nordlander. "Geometric dependence of the line width of localized surface plasmon resonances." *The Journal of Physical Chemistry Letters* 4, no. 8 (2013): 1352–1357. doi:10.1021/jz4004137.
13. Mulvaney, Paul. "Surface plasmon spectroscopy of nanosized metal particles." *Langmuir* 12, no. 3 (1996): 788–800. doi:10.1021/la9502711.
14. Schwartzberg, Adam M., and Jin Z. Zhang. "Novel optical properties and emerging applications of metal nanostructures." *The Journal of Physical Chemistry C* 112, no. 28 (2008): 10323–10337. doi:10.1021/jp801770w.
15. Li, Ming, Scott K. Cushing, Jianming Zhang, Jessica Lankford, Zoraida P. Aguilar, Dongling Ma, and Nianqiang Wu. "Shape-dependent surface-enhanced Raman scattering in gold–Raman-probe–silica sandwiched nanoparticles for biocompatible applications." *Nanotechnology* 23, no. 11 (2012): 115501. doi:10.1088/0957-4484/23/11/115501.
16. Willets, Katherine A., and Richard P. Van Duyne. "Localized surface plasmon resonance spectroscopy and sensing." *Annual Review of Physical Chemistry* 58 (2007): 267–297. doi:10.1146/annurev.physchem.58.032806.104607.
17. Lee, Kyeong-Seok, and Mostafa A. El-Sayed. "Dependence of the enhanced optical scattering efficiency relative to that of absorption for gold metal nanorods on aspect ratio, size, end-cap shape, and medium refractive index." *The Journal of Physical Chemistry B* 109, no. 43 (2005): 20331–20338. doi:10.1021/jp054385p.
18. Nallathamby, Prakash D., Tao Huang, and Xiao-Hong Nancy Xu. "Design and characterization of optical nanorulers of single nanoparticles using optical microscopy and spectroscopy." *Nanoscale* 2, no. 9 (2010): 1715–1722. doi:10.1039/C0NR00303D.
19. Tong, Ling, Qingshan Wei, Alexander Wei, and Ji-Xin Cheng. "Gold nanorods as contrast agents for biological imaging: Optical properties, surface conjugation and photothermal effects." *Photochemistry and Photobiology* 85, no. 1 (2009): 21–32. doi:10.1111/j.1751-1097.2008.00507.x.
20. Haes, Amanda J., Christy L. Haynes, Adam D. McFarland, George C. Schatz, Richard P. Van Duyne, and Shengli Zou. "Plasmonic materials for surface-enhanced sensing and spectroscopy." *MRS Bulletin* 30, no. 5 (2005): 368–375. doi:10.1557/mrs2005.100.
21. Li, Ming, Honglei Gou, Israa Al-Ogaidi, and Nianqiang Wu. "Nanostructured sensors for detection of heavy metals: A review." *ACS Sustainable Chemistry & Engineering* 1, no. 7 (2013): 713–723. doi:10.1021/sc400019a.
22. Wang, Hui, Carly S. Levin, and Naomi J. Halas. "Nanosphere arrays with controlled sub-10-nm gaps as surface-enhanced Raman spectroscopy substrates." *Journal of the American Chemical Society* 127, no. 43 (2005): 14992–14993. doi:10.1021/ja055633y.
23. Gryczynski, Ignacy, Joanna Malicka, Zygmunt Gryczynski, and Joseph R. Lakowicz. "Radiative decay engineering 4. Experimental studies of surface plasmon-coupled directional emission." *Analytical Biochemistry* 324, no. 2 (2004): 170–182. doi:10.1016/j.ab.2003.09.036.
24. Lakowicz, Joseph R. "Radiative decay engineering 5: Metal-enhanced fluorescence and plasmon emission." *Analytical Biochemistry* 337, no. 2 (2005): 171–194. doi:10.1016/j.ab.2004.11.026.
25. Sönnichsen, C., T. Franzl, Tv Wilk, Gero von Plessen, J. Feldmann, O. V. Wilson, and Paul Mulvaney. "Drastic reduction of plasmon damping in gold nanorods." *Physical Review Letters* 88, no. 7 (2002): 077402. doi:10.1103/PhysRevLett.88.077402.
26. Prodan, Emil, Corey Radloff, Naomi J. Halas, and Peter Nordlander. "A hybridization model for the plasmon response of complex nanostructures." *Science* 302, no. 5644 (2003): 419–422. doi:10.1126/science.1089171.

27. Passarelli, Nicolás, Luis A. Pérez, and Eduardo A. Coronado. "Plasmonic interactions: From molecular plasmonics and fano resonances to ferroplasmons." *ACS Nano* 8, no. 10 (2014): 9723–9728. doi:10.1021/nn505145v.

28. Kelly, K. Lance, Eduardo Coronado, Lin Lin Zhao, and George C. Schatz. "The optical properties of metal nanoparticles: The influence of size, shape, and dielectric environment." (2003): 668–677. doi:10.1021/jp026731y.

29. Tokel, Onur, Fatih Inci, and Utkan Demirci. "Advances in plasmonic technologies for point of care applications." *Chemical Reviews* 114, no. 11 (2014): 5728–5752. doi:10.1021/cr4000623.

30. Wang, Shaopeng, Xiaonan Shan, Urmez Patel, Xinping Huang, Jin Lu, Jinghong Li, and Nongjian Tao. "Label-free imaging, detection, and mass measurement of single viruses by surface plasmon resonance." *Proceedings of the National Academy of Sciences* 107, no. 37 (2010): 16028–16032. doi:10.1073/pnas.1005264107.

31. Masson, Jean-Francois. "Surface plasmon resonance clinical biosensors for medical diagnostics." *ACS Sensors* 2, no. 1 (2017): 16–30. doi:10.1021/acssensors.6b00763.

32. Paterson, S., and R. De La Rica. "Solution-based nanosensors for in-field detection with the naked eye." *Analyst* 140, no. 10 (2015): 3308–3317. doi:10.1039/C4AN02297A.

33. Peng, Chifang, Xiaohui Duan, Gabriel Wafala Khamba, and Zhengjun Xie. "Highly sensitive "signal on" plasmonic ELISA for small molecules by the naked eye." *Analytical Methods* 6, no. 24 (2014): 9616–9621. doi:10.1039/C4AY01993H.

34. Elghanian, Robert, James J. Storhoff, Robert C. Mucic, Robert L. Letsinger, and Chad A. Mirkin. "Selective colorimetric detection of polynucleotides based on the distance-dependent optical properties of gold nanoparticles." *Science* 277, no. 5329 (1997): 1078–1081. doi:10.1126/science.277.5329.1078.

35. Saha, Krishnendu, Sarit S. Agasti, Chaekyu Kim, Xiaoning Li, and Vincent M. Rotello. "Gold nanoparticles in chemical and biological sensing." *Chemical Reviews* 112, no. 5 (2012): 2739–2779. doi:10.1021/cr2001178.

36. Howes, Philip D., Subinoy Rana, and Molly M. Stevens. "Plasmonic nanomaterials for biodiagnostics." *Chemical Society Reviews* 43, no. 11 (2014): 3835–3853. doi:10.1039/C3CS60346F.

37. Lim, Wan Qi, and Zhiqiang Gao. "Plasmonic nanoparticles in biomedicine." *Nano Today* 11, no. 2 (2016): 168–188. doi:10.1016/j.nantod.2016.02.002.

38. Zhou, Wen, Xia Gao, Dingbin Liu, and Xiaoyuan Chen. "Gold nanoparticles for in vitro diagnostics." *Chemical Reviews* 115, no. 19 (2015): 10575–10636. doi:10.1021/acs.chemrev.5b00100.

39. Wang, Meijia, Chunyan Sun, Lianying Wang, Xiaohui Ji, Yubai Bai, Tiejin Li, and Jinghong Li. "Electrochemical detection of DNA immobilized on gold colloid particles modified self-assembled monolayer electrode with silver nanoparticle label." *Journal of Pharmaceutical and Biomedical Analysis* 33, no. 5 (2003): 1117–1125. doi:10.1016/S0731-7085(03)00411-4.

40. Verma, Mohit S., Jacob L. Rogowski, Lyndon Jones, and Frank X. Gu. "Colorimetric biosensing of pathogens using gold nanoparticles." *Biotechnology Advances* 33, no. 6 (2015): 666–680. doi:10.1016/j.biotechadv.2015.03.003.

41. Boles, Michael A., Michael Engel, and Dmitri V. Talapin. "Self-assembly of colloidal nanocrystals: From intricate structures to functional materials." *Chemical Reviews* 116, no. 18 (2016): 11220–11289. doi:10.1021/acs.chemrev.6b00196.

42. Borges, Joao, and Joao F. Mano. "Molecular interactions driving the layer-by-layer assembly of multilayers." *Chemical Reviews* 114, no. 18 (2014): 8883–8942. doi:10.1021/cr400531v.

43. Jones, Matthew R., Kyle D. Osberg, Robert J. Macfarlane, Mark R. Langille, and Chad A. Mirkin. "Templated techniques for the synthesis and assembly of plasmonic nanostructures." *Chemical Reviews* 111, no. 6 (2011): 3736–3827. doi:10.1021/cr1004452.

44. Vogel, Nicolas, Markus Retsch, Charles-Andre Fustin, Aranzazu Del Campo, and Ulrich Jonas. "Advances in colloidal assembly: The design of structure and hierarchy in two and three dimensions." *Chemical Reviews* 115, no. 13 (2015): 6265–6311. doi:10.1021/cr400081d.

45. Walther, Andreas, and Axel H. E. Muller. "Janus particles: Synthesis, self-assembly, physical properties, and applications." *Chemical Reviews* 113, no. 7 (2013): 5194–5261. doi:10.1021/cr300089t.

46. Valentini, Paola, Roberto Fiammengo, Stefania Sabella, Manuela Gariboldi, Gabriele Maiorano, Roberto Cingolani, and Pier Paolo Pompa. "Gold-nanoparticle-based colorimetric discrimination of cancer-related point mutations with picomolar sensitivity." *ACS Nano* 7, no. 6 (2013): 5530–5538. doi:10.1021/nn401757w.

47. Xianyu, Yunlei, Zhuo Wang, and Xingyu Jiang. "A plasmonic nanosensor for immunoassay via enzyme-triggered click chemistry." *ACS Nano* 8, no. 12 (2014): 12741–12747. doi:10.1021/nn505857g.

48. Xianyu, Yunlei, Yiping Chen, and Xingyu Jiang. "Horseradish peroxidase-mediated, iodide-catalyzed cascade reaction for plasmonic immunoassays." *Analytical Chemistry* 87, no. 21 (2015): 10688–10692. doi:10.1021/acs.analchem.5b03522.

49. Liu, Dingbin, Zhantong Wang, Albert Jin, Xinglu Huang, Xiaolian Sun, Fu Wang, Qiang Yan et al. "Acetylcholinesterase-catalyzed hydrolysis allows ultrasensitive detection of pathogens with the naked eye." *Angewandte Chemie* 125, no. 52 (2013): 14315–14319. doi:10.1002/ange.201307952.

50. Valentini, Paola, and Pier Paolo Pompa. "A universal polymerase chain reaction developer." *Angewandte Chemie International Edition* 55, no. 6 (2016): 2157–2160. doi:10.1002/anie.201511010.

51. Nie, Xin-Min, Rong Huang, Cai-Xia Dong, Li-Juan Tang, Rong Gui, and Jian-Hui Jiang. "Plasmonic ELISA for the ultrasensitive detection of *Treponema pallidum*." *Biosensors and Bioelectronics* 58 (2014): 314–319. doi:10.1016/j.bios.2014.03.007.

52. Wang, Fangfang, Shuzhen Liu, Mingxia Lin, Xing Chen, Shiru Lin, Xiazhen Du, He Li et al. "Colorimetric detection of microcystin-LR based on disassembly of orient-aggregated gold nanoparticle dimers." *Biosensors and Bioelectronics* 68 (2015): 475–480. doi:10.1016/j.bios.2015.01.037.

53. Wei, Luming, Xiaoying Wang, Chao Li, Xiaoxi Li, Yongmei Yin, and Genxi Li. "Colorimetric assay for protein detection based on 'nano-pumpkin' induced aggregation of peptide-decorated gold nanoparticles." *Biosensors and Bioelectronics* 71 (2015): 348–352. doi:10.1016/j.bios.2015.04.072.

54. Chandrawati, Rona, and Molly M. Stevens. "Controlled assembly of peptide-functionalized gold nanoparticles for label-free detection of blood coagulation Factor XIII activity." *Chemical Communications* 50, no. 41 (2014): 5431–5434. doi:10.1039/C4CC00572D.

55. Wu, Shaojue, Si Yu Tan, Chung Yen Ang, Zhong Luo, and Yanli Zhao. "Oxidation-triggered aggregation of gold nanoparticles for naked-eye detection of hydrogen peroxide." *Chemical Communications* 52, no. 17 (2016): 3508–3511. doi:10.1039/C5CC09447J.

56. Richards, Sarah-Jane, Elizabeth Fullam, Gurdyal S. Besra, and Matthew I. Gibson. "Discrimination between bacterial phenotypes using glyco-nanoparticles and the impact of polymer coating on detection readouts." *Journal of Materials Chemistry B* 2, no. 11 (2014): 1490–1498. doi:10.1039/C3TB21821J.

57. Guo, Longhua, Yang Xu, Abdul Rahim Ferhan, Guonan Chen, and Dong-Hwan Kim. "Oriented gold nanoparticle aggregation for colorimetric sensors with surprisingly high analytical figures of merit." *Journal of the American Chemical Society* 135, no. 33 (2013): 12338–12345. doi:10.1021/ja405371g.

58. Bui, Minh-Phuong Ngoc, Snober Ahmed, and Abdennour Abbas. "Single-digit pathogen and attomolar detection with the naked eye using liposome-amplified plasmonic immunoassay." *Nano Letters* 15, no. 9 (2015): 6239–6246. doi:10.1021/acs.nanolett.5b02837.

59. Gormley, Adam J., Robert Chapman, and Molly M. Stevens. "Polymerization amplified detection for nanoparticle-based biosensing." *Nano Letters* 14, no. 11 (2014): 6368–6373. doi:10.1021/nl502840h.

60. Aili, Daniel, Robert Selegård, Lars Baltzer, Karin Enander, and Bo Liedberg. "Colorimetric protein sensing by controlled assembly of gold nanoparticles functionalized with synthetic receptors." *Small* 5, no. 21 (2009): 2445–2452. doi:10.1002/smll.200900530.

61. Chen, Yiping, Yunlei Xianyu, Jing Wu, Binfeng Yin, and Xingyu Jiang. "Click chemistry-mediated nanosensors for biochemical assays." *Theranostics* 6, no. 7 (2016): 969. doi:10.7150/thno.14856.

62. Wei, W. Yu, and Ian M. White. "Inkjet-printed paper-based SERS dipsticks and swabs for trace chemical detection." *Analyst* 138, no. 4 (2013): 1020–1025. doi:10.1039/C2AN36116G.

63. Lee, Chang H., Limei Tian, and Srikanth Singamaneni. "SERS based swab for rapid trace detection on real-world surfaces." *ACS Applied Materials & Interfaces* 2, no. 12 (2010): 3429–3435. doi:10.1021/am1009875.

64. Lee, Chang H., Mikella E. Hankus, Limei Tian, Paul M. Pellegrino, and Srikanth Singamaneni. "Highly sensitive surface-enhanced Raman scattering substrates based on filter paper loaded with plasmonic nanostructures." *Analytical Chemistry* 83, no. 23 (2011): 8953–8958. doi:10.1021/ac2016882.

65. Chang, Shih-Hsin, James Nyagilo, Jiaqi Wu, Yaowu Hao, and Digant P. Davé. "Optical fiber-based surface-enhanced Raman scattering sensor using Au nanovoid arrays." *Plasmonics* 7, no. 3 (2012): 501–508. doi:10.1007/s11468-012-9335-7.

66. Guarrotxena, Nekane, and Guillermo C. Bazan. "Antitags: SERS-encoded nanoparticle assemblies that enable single-spot multiplex protein detection." *Advanced Materials* 26, no. 12 (2014): 1941–1946. doi:10.1002/adma.201304107.

67. Lee, Hiang Kwee, Yih Hong Lee, In Yee Phang, Jiaqi Wei, Yue-E. Miao, Tianxi Liu, and Xing Yi Ling. "Plasmonic liquid marbles: A miniature substrate-less SERS platform for quantitative and multiplex ultratrace molecular detection." *Angewandte Chemie International Edition* 53, no. 20 (2014): 5054–5058. doi:10.1002/anie.201401026.

68. Li, Ming, Jing Lu, Ji Qi, Fusheng Zhao, Jianbo Zeng, Jorn Chi-Chung Yu, and Wei-Chuan Shih. "Stamping surface-enhanced Raman spectroscopy for label-free, multiplexed, molecular sensing and imaging." *Journal of Biomedical Optics* 19, no. 5 (2014): 050501. doi:10.1117/1.JBO.19.5.050501.

69. Zheng, Jing, Yaping Hu, Junhui Bai, Cheng Ma, Jishan Li, Yinhui Li, Muling Shi, Weihong Tan, and Ronghua Yang. "Universal surface-enhanced Raman scattering amplification detector for ultrasensitive detection of multiple target analytes." *Analytical Chemistry* 86, no. 4 (2014): 2205–2212. doi:10.1021/ac404004m.

2 Plasmonic Nanosensors
Classification, Properties, Applications, and Future Perspectives

2.1 INTRODUCTION

The field of plasmonics is bourgeoning at the interface of nanotechnology, photonics, and electronics [1]. Plasmonics is a burgeoning field of the study showing tangible signs of progress in nanomaterials' development and nanophotonics and nanooptics applications [2–5]. By the light coupling with nano-objects at the nanoscale, manufacturing various novel optical properties occurs using plasmonics. The collective motion of conduction electrons is exploited thru these optical properties in plasmons (metals). The plasmons excitation is escorted through localization and theatrical augmentation of the electric field allied by light at optical frequencies.

The plasmonics of nanomaterials are based on surface plasmon (SP) oscillations. It has found the leading discovered platforms for chemical sensing. It is defined as metallic nanostructures because of the sensitivity of SPR, the basis of the sensors formed in the metal's locality surface to the dielectric medium [6]. For example, various metals, such as Cu, Ag, and Au, support SP waves in the visible range and NIR. The plasmonic materials of these metals in chemical sensors offer many applications. For example, small sensing area [7], simplistic surface chemistry to immobilize molecular sensing elements [8], simple incorporation with microfluidics [9], and corresponding to excellent sensitivity, small device footprint, and the capability to detect diverse types of chemical pollutants [10]. All these promising properties in the development of various SPR-based sensors justify the proposed elevated research activity.

2.2 TYPES OF PLASMON MODES

The propagating surface plasmons (PSPs) and localized surface plasmons (LSPs) are two types of plasmon modes.

2.2.1 LOCALIZED SURFACE PLASMONS

The LSP includes the collective oscillations of free electrons and the associated oscillation of the EM field in MNPs. The local composition, optical environment,

DOI: 10.1201/9781003128281-2

shape, and particle size are pivotal to the resonance frequency [11,12]. The resonance frequency characteristically lies in the visible NIR part of the spectrum for the Cu, Ag, and Au noble metals' nanostructures. Initially, in various applications, LSPs have appeared as a fascinating substitute for PSPs because of translational symmetry and momentum matching deficiency. However, LSPs usually have a larger spectral width than PSPs, for example, Au nanostructures in contrast to a spectral width of ~50 nm for PSP, the resonance width for LSPs characteristically surpasses ~80–100 nm full width at half maximum (FWHM). For many applications, this augmented width knowingly confines the competence of LSPs [13–15].

2.2.2 PROPAGATING SURFACE PLASMONS

Because of their significance for biorecognition/biosensing submissions, PSPs have been widely studied and fascinated by many research groups' attention. PSPs might be surface electromagnetic waves and are reinforced at a dielectric/metal interface. PSPs encompass an EM wave because of binding with the collective motion of mobile charges on the metal surface. Because of this coherent interaction compared to a free photon, PSP contributed greater momentum of the same frequency. The nanohole array couplers are a significant example of various sorts of momentum matching methods used for the excitation of PSPs, that is, grating coupling [16], nanohole array couplers [17], and prism coupling [18]. The Turbadar–Kretschmann–Raether prism geometry is a prominent example of PSP excitation, which is also referred to as SPR [18]. In this method, after deposition on one of the prism's faces, polarized light is reflected from a thin (~50 nm) metal film when directed into a glass prism. The excitation of PSPs is escorted in the reflectance spectrum by a minimum for a fixed angle of incidence.

2.3 CLASSIFICATION OF THE PLASMONIC NANOMATERIALS

Nanomaterials can be classified in several ways. This usually depends on the application and the philosophy adopted. Based on their appearance, nanomaterials can be classified as either natural or engineered. There are various engineered nanomaterials, and a variety of others are anticipated to be developed. They can be in nanoscale, that is, quantum dots (0D), surface films (one-dimensional (1D)), fibers (two-dimensional (2D)), and particles (three-dimensional (3D)). They could occur in various systems, that is, single fuses, aggregated with spherical, tubular, and irregular shapes. In nanotechnology, nanomaterials have applications and possess unique chemical and physical properties from normal chemicals. Here, we have described nanomaterials based on the dimensions and structural configuration [19].

2.3.1 BASED ON DIMENSIONS

Based on the dimensions, nanomaterials can be classified into four categories.

Zero-dimensional (0D): A 0D nanomaterial exhibits electrons' confined movement in all three dimensions. Most of these nanomaterials are spherical, with a narrow diameter ranging from 1 to 10 nm. The most common example of 0D materials is nanospheres or a quantum dot.

One-dimensional (1D): This type of nanomaterials has electrons' movement in one of their directions (X plane). This type of nanomaterial has one dimension of nanostructures out from the nanometer range. This leads to the needle-like shape of nanomaterials, such as nanowires and nanorods.

Two-dimensional (2D): The electrons can move freely in two of their directions (X–Y plane). This type of nanomaterials has two dimensions out of the nanometer range. Examples are different nanofilms, nanowells, and nanosheets.

Three-dimensional (3D): In this type of nanomaterials, free electrons can move easily in the X, Y, and Z planes. These materials partake in all the dimensions out of the nanometer range. This type of materials comprises the individual blocks in the 1–100 nm range (nanometer), bulk materials. This type of nanomaterials forms interfaces, and the 0D, 1D, 2D structural elements are near each other, which comprise fibrous, powders, polycrystalline, and multilayer materials. Therefore, it is also known as a hierarchical structure.

2.3.2 BASED ON THE STRUCTURAL CONFIGURATION

Based on the structures, the nanomaterials can be classified into the following categories.

Carbon-based nanomaterials: Taking the forms of ellipsoids, tubes, and hollow spheres, these nanomaterials are generally made of carbon. Sheets and cylindrical carbon nanomaterials are called graphene and carbon nanotubes (CNTs), respectively, while ellipsoids and circular carbon nanomaterials are known as fullerene.

Metal-based nanomaterials: Metal is the most important component of these materials, which comprises quantum dots, nanogold, metal oxides, for example, titanium dioxide (TiO_2) and nanosilver.

Dendrimers: These nanomaterials are polymers in the nano-ranged size made up of branched units. The surface of these materials can be altered to execute precise functions, which have numerous chains of the dendrimers.

Composites: Nanocomposites are those materials whose at least one phase has one, two, or three dimensions in nanoscale and are described as multiphase solid material. Copolymers, gels, and colloids are the most general examples of these materials. The schematic representation of dissimilar sorts of plasmonic nanomaterials is shown in Figure 2.1.

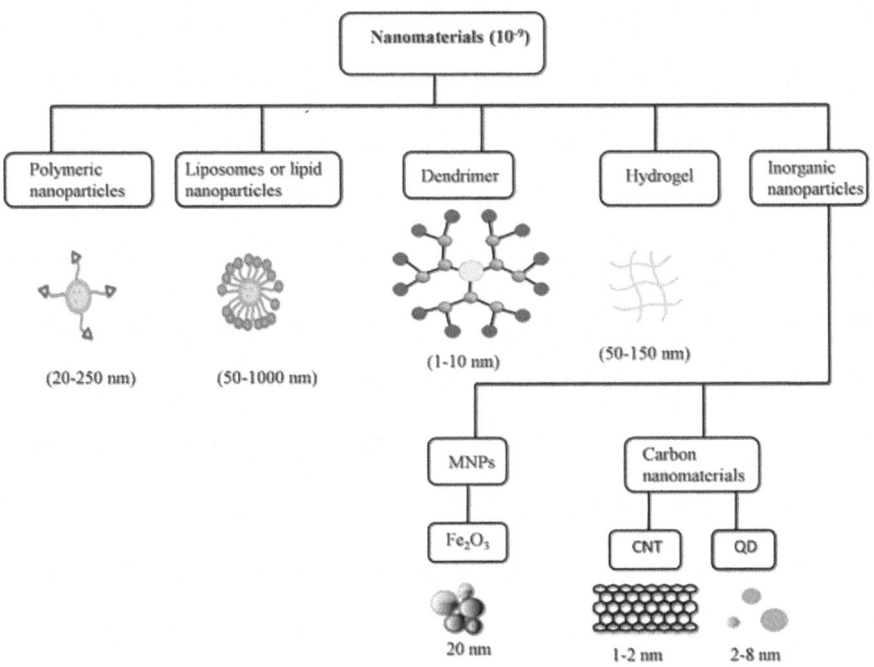

FIGURE 2.1   Classification of plasmonic nanomaterials. (Reprinted with permission from Ref. [4]. Copyright 2019 The Springer Nature.)

2.4   PLASMONIC NPs

The size of particles, interparticle distance, and the NPs' shape are the key applications of plasmonic NPs. Since then, on the properties of NPs, modification, and synthesis, a vast amount of work has been published. We must know about the nucleation process and growth of NPs to control the NPs' size distribution. The NPs can be synthesized in various sizes, shapes, and structures by utilizing multiple methods. Lamer and Dinegar show that the formation of monodisperse colloids involves the slower growth of the nuclei, followed by rapid nucleation [20]. During the nucleation stage, thru quick injection of the reducing agent, the fast nucleation stage can be set up; tiny seed or monomer forms a massive amount due to the vigorous reaction. The monomer concentration rises and reaches above the supersaturation concentration as the reaction proceeds. Nucleation occurs, and many nuclei form in a short burst at this stage. No further nucleation will occur until the monomer concentration remains lower than the supersaturation concentration. The reaction then enters the growth stage, in which particles grow by molecular addition continuously.

In contrast to larger particles, smaller particles grow faster. Throughout the size-focusing process, thru stopping the reaction at a suitable time, narrow size distribution and desired size can be achieved. In terms of diameter, NPs are classified into ultrafine, fine, and coarse particles.

2.5 PROPERTIES OF THE NPs

The materials show unusual and new mechanical, physical, and chemical properties, as they are abridged in their size from bulk to nanoscale. Noble MNPs have gained significant scientific interest owing to these properties. Some of these properties are described in the following sections.

2.5.1 SURFACE PLASMON RESONANCE

Plasmons are the quantized wave in the collective oscillation of the mobile electrons in metals generated when many electrons get disturbed from their equilibrium position [21]. Metal surrounded by the plasma of free electrons can be considered positively charged atom nuclei. As electrons are particles carrying a negative charge, they generate an electric field during vibration. When this electric field resonates with the external electric field, the resulting phenomenon is called SPR (Figure 2.2).

On the NPs' surface, the collective oscillation of the electron's plasma when the NPs are irradiated with light is called SPR [22]. The light is absorbed, and SPR resonance arises when an incoming EM wave matches the electron cloud oscillations. However, when the SPR is localized to volume with dimensions, it is known as LSPR [23]. When the size of a particle is small to support the propagating wave, light interacts with the metal particles to generate LSPR, smaller than the wavelength of light. The wavelength corresponding to LSPR depends on the dielectric environment, material composition, extent of aggregation and shape, size of MNPs [24]. The SPR mechanism—the atoms present in the metal are very close to each other. They can also share electrons (electron delocalization). A bundle of energy carried by electromagnetic radiation (EMR) is called photons. EMR can impart some of its energy to the particles [25].

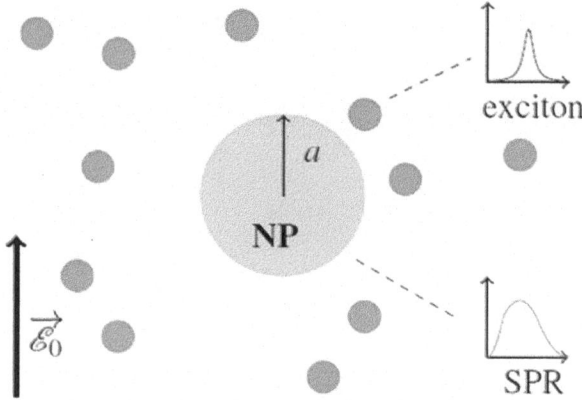

FIGURE 2.2 Schematic presentation of SPR where, because of strong coupling with incident light, the free conduction electrons of MNPs are driven into oscillation. (Reprinted with permission from Ref. [5]. Copyright 2018 American Chemical Society.)

The electrons moved away from their equilibrium position because of the disturbed charged electron cloud by interacting with the EMR or electrons. The kinetic energy of electrons corresponds to reemergence to their original equilibrium position. This kinetic energy exceeds its equilibrium mark and starts oscillating back and forth [26]. The plasmons result from these collective oscillations of the conduction electrons. As it involved the electrons present on the surface in the oscillations, this collective oscillation is called SPR [27]. SPR is a surface phenomenon contingent on many factors such as shape, size, dielectric property of the metal, and the surrounding medium [28]. The plasma frequency lies in the spectrum's UV region for many metals, for example, Hg, Pb, Cd, where the NPs do not show any color.

In Cu, Au, and Ag, because of the d–d transitions, the plasma frequency lies in the visible region of the spectrum. The entire charge distribution of particles oscillates because of the excitation of a dipole plasmon resonance at the frequency of the incident electric field. The NPs with small dimensions have a sharp and single absorption band. For instance, the intense (ruby red) color of AuNPs is attributed to the resonance condition at the visible wavelengths. The SPR maximum of the AuNPs lies between 520 and 550 nm. Therefore, a redshift occurs in their plasmon resonance by increasing the particles' size, changing the color of their corresponding solutions.

Anisotropic NPs arise in various SPR modes. Simultaneously, nanorods can be used as a typical example for demonstrating the dependency of optical properties on the particle's shape. The nanorods' dipole resonance split into longitudinal and transverse resonances, owing to the diverse dimensions along the particles' length and width. The bands' positions are contingent upon the absolute dimensions and the aspect ratio of the particles [29]. The resonance at the longer wavelength (longitudinal plasmon resonance) corresponds with the oscillations along the nanorods' length. In contrast, the oscillations along the nanorods' width result from the resonance at the shorter wavelength (transverse plasmon resonance).

The aspect ratio is the basis of SPR of the nanorods. For instance, various positions of the two bands are shown by the gold nanorods (GNRs) of different aspect ratios. It will couple the surface plasmons of nanomaterials. Their oscillating electric fields interact to give new resonance because of the high sensitivity to the proximity of the SPR frequency of other particles when the two or more discrete nanomaterials are near each other [30]. A shift in the SPR peak of AuNPs occurs toward the longer wavelength, resulting in the aggregation of peak broadening. Because of the interparticle plasmon coupling, the change in the color of the aggregated NPs occurs. This property of nanomaterials makes them an alluring contender toward the colorimetric sensor.

In nanomaterials and colorimetric sensing, before analyzing nanomaterials' colorimetric visual sensing application, the basics of sensors should be discussed first. The conversion of physical or chemical changes into a signal can be observable by an observer or instrument called a sensor. Effective sensors must have the following characteristics:

- the capability to detect or analyze the specific target species despite other chemical species;
- it essential to detect the target analyte up to the minimum signal;
- the sensor should keep its features when used irrespective of the changing conditions and environments;
- for a certain set of conditions throughout its lifetime, it is a key feature that the sensor can be trusted to give the same output/signal value;
- it must give the correct concentration of the target ions to be sensed, and
- up to the desired range, the sensor must linearly respond to the analyte concentration.

Thus, the detection of the analyte using the colorimetric method can be easily observed through the naked eyes and transformed into a color change. This is mainly the molecular detection method. Based on the size, shape, aggregation behavior, and surrounding environment, MNPs display emit bright resonance light scattering and eclectic colors of various wavelengths. Thus, spherical and non-spherical (rods and prisms) with unique size-dependent optical properties can be widely used for the colorimetric sensing of different analytes, for instance, metal cations, proteins, cells, nucleic acid, etc. Silver and gold nanostructures are most alluring for optical purposes because of their outstanding optical properties [31,32]. As described above, the colorimetric methods based on NPs depend on the alteration in the optical properties because of the morphology transition, aggregations, which thus correspond to the distinctive color change.

2.5.2 SURFACE-TO-VOLUME RATIO

At the surface of the material, chemical reactions always take place. The surface-to-volume ratio has a substantial effect on various properties of the materials. The reactivity will be higher with a higher surface for the same volume.

- The surface area of a sphere $= 4\pi r^2$
- The volume of the sphere $= 4/3 \ (\pi r^3)$
- Surface area to the volume ratio will be $4\pi r^2/\{4/3(\pi r^3)\} - 3/r$

However, with a decrease in the sphere's size, concomitant surface-to-volume ratio increases. Their surface-to-volume ratio increases dramatically with small-sized particles. For example, a sugar cube reacting with water will dissolve the outside of the sugar. If the same sugar cube is cut into various pieces, each cut will make new outer surfaces for the water to dissolve. The same volume of the sugar will have more surface area for the smaller pieces of the sugar. A more significant number of reaction sites partake by a particle having a high surface area than the particles with a low surface area, resulting in higher chemical reactivity. Small particles react faster with a more surface-to-volume ratio because of the more surface area, which provides more reaction sites, leading to more chemical reactivity [33].

2.5.3 Reactivity

Just as the nanomaterials' electronic and optical possessions, the changes in the chemical reactivity can also be foreknown, numerous factors affect the reactivity depends, for example:

- Shape and size of the inorganic core;
- Interaction with the ions;
- Functionalization;
- Surface chemistry, and
- The cooperative effect with the other metal ions presents in the solution.

In bulk materials, surface atoms contribute only a minor portion of the total number of atoms. In the NPs, the radii's curvature is so high that all the atoms are present at the surface and lie close to each other. The small particles possess all surface atoms, and these surface atoms have a lower coordination number compared to their bulk counterparts. The reactivity of the single atoms increases with the decrease in coordination. Thus, with the decrease in particle size, the reactivity of the nanoscale materials increases. For instance, bulk gold is a noble metal. When it reaches its nanoscale, it acts as a good catalyst for organic reactions [34].

2.5.4 Quantum Confinement

Drastic changes in the nanomaterials' vibrational and electronic properties can be observed when the materials' macroscopic dimensions are continuously reduced to the nanometer size. In contrast to the electron's wavelength, when the size of particles is small, this effect is observed. The atomic realm of the particles is called a quantum, and confinement shows the confined motion of the arbitrarily oscillating electrons to the specific energy levels. The decrease in the particles' dimensions makes the discrete energy levels, which widens the bandgap, increasing the band energy gap. As the bandgap and wavelength are inversely related, the wavelength decreases with a decrease in size [35]. The smaller is the dimensions, the greater is the bandgap. It drastically changed most bulk semiconductor properties because of this increase in the bandgap, such as phonon propagation, nonlinear and linear optical properties, and electronic band structure.

2.5.5 Melting Point

The melting point of the bulk counterparts is independent of its size, but in the nanoscale materials, with the decreased size, a depression in the melting point is observed. The melting temperature depends on the size of the NPs. According to Pawlow, where a liquid particle and solid particle are in equilibrium with their vapors having similar mass is known as melting temperature of NPs. The change in the melting point was also observed, which shows that it is inversely proportional to the particle radius [36]. Because of the thermal properties, change in thermodynamic properties, increased surface-to-volume ratio of nanoscale materials than bulk materials, and changes in melting point occur. Depression in melting point can be observed in NPs, nanotubes, etc. At the surface of NPs, a substantial number of atoms are present owing to a rise in the surface-to-volume ratio. Fewer neighboring atoms surround these atoms and thus feel less cohesive energy. Cohesive energy is required to free the atom from the surface, which measures thermal energy. According to Lindemann's criterion, the melting temperature of a nanomaterial is directly proportional to its cohesive energy [37]. The surface atoms require less energy because they have reduced cohesive energy and fewer bonds. NPs' melting point decreases with the particle reaching the critical diameter (less than 50 nm) for common manufacturing metals.

2.6 APPLICATIONS OF PLASMONIC SENSORS

2.6.1 CHIP-BASED PLASMONIC SENSORS

The alterations in the neighboring medium's refractive index, planar plasmonic substrates, can be developed for plasmonic sensing support propagating mixed SPP/LSPR mode or SPP mode. To supply the additional momentum required to match free-space light, SPP can only be thrilled thru a grating or a prism. The time-determined and angle reflectivity measurements enable the classical setup of chip-based SPP sensors of a noble metal film's SPP mode [38–42]. Because of the higher refractive index of the molecules on the plasmonic metal film compared to the aqueous solution, the SPP band works as a sensor and redshifts when the analyte (molecules) binds to immobilized ligands [43]. Consistent with the Drude model, with the wavelength of the SPR peak, the refractive index of the neighboring medium fluctuates linearly [44]. Therefore, the refractive index sensitivity ($S = \Delta\lambda_p/\Delta n$) is shown in units of nm per RIU, where Δn is the refractive index of the surrounding medium and $\Delta\lambda_p$ is the plasmon frequency of the metal. A figure of merit (FOM = S/FWHM) can appraise the sensing ability of a plasmonic sensor.

Here, FWHM stands for the full width at half-maximum of the width of SPR peak, and S stands for refractive index sensitivity. The overlapping of the SPR peak of the metal film with the absorption band of the chromophore is the only method used to progress the refractive index sensitivity of a plasmonic sensor that binds to the plasmonic metal [45,46]. As an alternative, the refractive index sensitivity can be enhanced by substituting the planar metal film with a large-area periodic nano-array pattern. In contrast to the planar films, the nano-arrays support a higher sensing area and stronger local EM field, such as a nano-triangle array, nano-disc array, or nanosphere array. When an analyte linker is present in the conjugated plasmonic NPs, the SPR peak relative to the Au film alone can also be coupled to an Au film in a sandwich configuration, which increases the shift of the SPR peak [47]. The manageable manufacture of large-area (cm²) plasmonic nano-array patterns is empowered with the fast development of nanolithography technique, which includes the applications of planar and colloidal substrates, which are summarized below.

In contrast to bulk materials, it is owing to the SPR band and tunable field distribution, miniaturized configuration, good repeatability, and facile integration with other components (e.g., microfluidics). In addition, a high density of "hot spots" with a more accessible accommodation into a portable analytical instrument ensuing in improved sensitivity in the order of a billion or more per cm^2.

Toward plasmonic sensing, the nanohole array is of great interest. As described earlier, corresponding to a very concentrated EM field, LSPR can occur in an individual nanohole near the edge [48]. SPP can be excited through the nanoholes at certain resonance wavelengths when the nanoholes are developed periodically over a large area [49]. The periodicity of the nanohole array is reliant on the SPP mode. SPP mode disappears for a non-periodic nanohole array, whereas the LSPR mode still exists. The LSPR and SPP modes can be sensed directly through incident light without a prism. This can occur where backside detection/front side illumination is possible, specifically in extraordinary optical transmission [48]. The periodic structures considerably simplify the configuration of SPP-based sensors by allowing a nanohole array-based plasmonic sensor. To have a small footprint for miniaturization and eliminating the typical angle-resolved reflection arrangement used in the Kretschmann configuration. The LSPR and SPP spectral peaks are highly sensitive toward dielectric properties at the interface in a periodic nanohole array with the adsorption of molecules [50,51]. Plasmonic nanohole array is conducive to multiplex detection [52], which enables integration into antibody-ligand binding kinetics, real-time measurements [53], and microfluidics [54]. Pang et al. described the successive inoculation of ethylene glycol into a water-based high-performance microfluidic nanohole array with a refractive index sensitivity of 1,520 nm per RIU [55]. Plasmonic chip-based sensors have three diverse applications [56]:

i. Long-range SPP (LRSPP) modes display orders of magnitude less damping than conventional SPP and can be easily fabricated [56].
ii. A prevailing tool to study the binding actions attained when the real-time measurement of reaction kinetics occurs in the plasmonic chip assimilated into a flow-cell [57].
iii. Label-free detection abolishes the gold standard for analytical assays. It uses many antibodies as employed in enzyme-linked immunosorbent assay (ELISA), making the operation and configuration simpler for the sensor [58].

The minor damping in LRSPP permits more than 1 mm of penetration depth, higher FOM, and narrower bandwidths [59], perfect for examining living beings [60]. When a thin metal film is implanted among two dielectrics, LRSPP modes are possible with similar refractive indices [61]. Figure 2.3 demonstrates the uses of plasmonic nanomaterials in several areas.

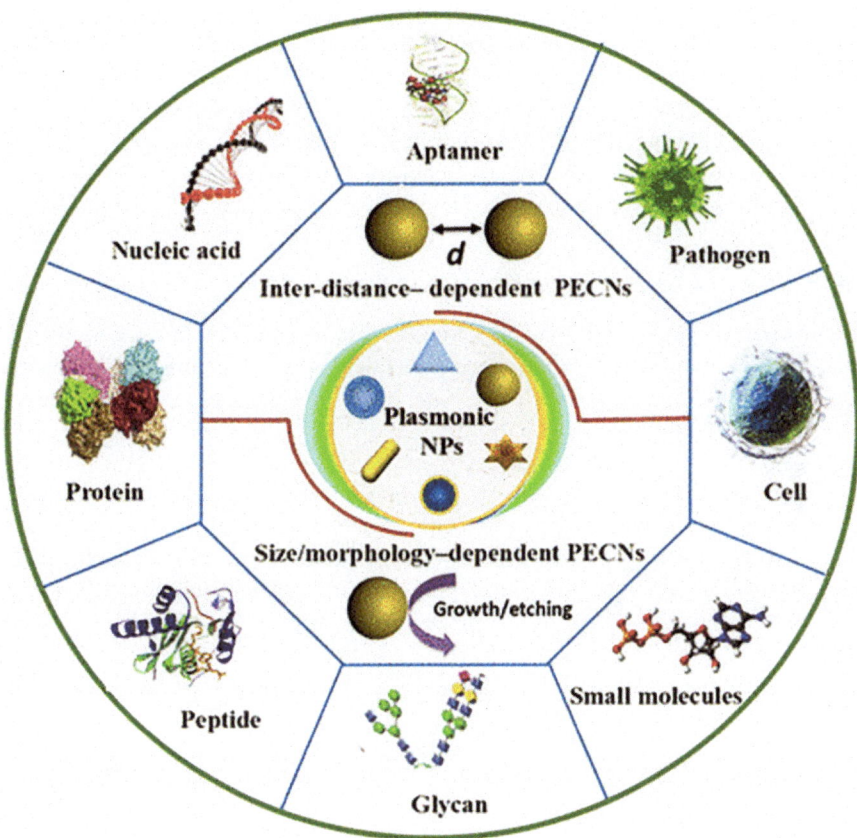

FIGURE 2.3 Schematic representation of the application of plasmon nanomaterials. (Reprinted with permission from Ref. [9]. Copyright 2017 American Chemical Society.)

2.6.2 NP-Based Colorimetric Sensors

The noble metal SPR rises, resonating with the incident EM radiation [62]. For instance, in the absorption spectrum, monodisperse AuNPs of wine-red color show a strong SPR peak with 20 nm in the aqueous solution. The color of the solution is changed when the AuNPs aggregate to some degrees, and then the position of the SPR peak will shift. The colorimetric sensors are manufactured based on this principle to detect numerous analytes in the aqueous media [63]. This type of sensor reduces the prices by providing rapid, visual, and direct detection of the analyte. The detection of heavy metals has been done by Hupp using colorimetric sensors [64]. To date, for the many heavy metal ion sensing, for example, Hg^{2+}, Pb^{2+}, As^{3+}, and Cu^{2+}, the colorimetric detection technique has been utilized [65]. Liu et al. used the oligonucleotide functionalized AuNPs for the colorimetric detection of Hg^{2+} ions [66]. The selectivity toward Hg^{2+} ions corresponds to the aggregation of AuNPs by the selective binding of T–T mismatches to the Hg^{2+} ions. Because of the coupling of SPR of the AuNPs nearby, the aggregation of AuNPs induces a color change. The developed sensor showed 3 µM of the limit of detection (LOD) for Hg^{2+} ions. However, the colorimetric detection method's LOD requires the preconcentration step. It is relatively high to detect trace metals that are rapid and simple. The applicability of colorimetric sensors remains a challenge for real-world sample matrices, such as plasma and human blood.

2.6.3 COLLOIDAL NP-BASED PLASMONIC SENSORS

To construct plasmonic sensors, NPs, such as Cu, Ag, and Au, have been utilized because they exhibit size- and shape-dependent scattering and LSPR absorption bands. Based on the LSPR peak shift, there are two categories of plasmonic sensors:

  i. the LSPR peak wavelength shifts when a target molecule binds to the surface of the NP, altering the local refractive index [67];
 ii. a shift of the LSPR and a color change occur when a target molecule brings the NPs into proximity, the plasmonic fields of multiple NPs are coupled [68].

Among various plasmonic NPs, colloidal AuNPs are the most used plasmonic transducer, owing to visible color change determined by the naked eye and their chemical stability. Through the visualization of change in color of the analyte concentration, it could be detected by the colorimetric detection method directly. Therefore, for detecting an analyte in a solution, colorimetric detection is a facile and straightforward technique. Functionalized AuNPs have been utilized in colorimetric detection of biomacromolecules, smaller analytes, and heavy metals [69]. Surface modification determines both the selectivity and sensitivity; hence, it is extremely important when designing a sensor. The colorimetric detection method is based on the distance-dependent LSPR coupling among AuNPs established by the Mirkin group in which polynucleotides were selectively detected [70].

In addition, using 11-mercaptoundecanoic acid (MUA) modified AuNPs, heavy metals, for example, Pb^{2+}, Cd^{2+}, and Hg^{2+} have been detected by the colorimetric technique [71]. Through an ion-templated chelation procedure, this modifier can detect divalent metal ions. The binding of a divalent metal ion with a carboxylic group causes aggregation, and an alteration in the extinction spectrum of the Au colloidal suspension occurs. Using DNA as the detection site, this method has been enhanced in specificity and sensitivity [72,73]. In this method, the change in color from red to blue of the solution occurs by adding metal ions forming DNA-linked aggregates because of the reversible dissociation/association of two complementary Au NPs functionalized with DNA. The mismatched T–T base pair selectively coordinated with the developed sensor when Hg^{2+} is present in the solution and allows quantification of the Hg^{2+} concentration [74]. Liu et al. reported on the DNA–Au NP-based sensing system to realize colorimetric detection of Hg^{2+}

at a room temperature using an optimal DNA sequence [75]. By replacing the thymidine with synthetic artificial bases, this process reports selective bind of the sensor to other metal ions [76]. The LSPR of NPs is extremely reliant on their size and shape in contrast to solid nanosphere. It is more sensitive to nanoshells and nanorods to changes in the local refractive index [77]. In contrast to a nanosphere counterpart, a longitudinal LSPR mode of nanorod displays six times advanced sensitivity [78]. The subsequent EM field is screened thru the refractive index and imagining how the concentrated charge, the dependence of refractive index sensitivity, can be explained conceptually on the nanostructure's geometry.

2.7 CONCLUSIONS, FUTURE PERSPECTIVES, AND CHALLENGES

The physics of plasmonics is interesting and rich. Recently, properties associated with LSPR, the noble MNPs fascinating the major research interest. Theoretical models have been well recognized with in-depth studies of LSPR and applied to various LSPR cases. For the applications and design in different fields, these theoretical advances have also led to photothermally induced optical trapping, SERS and optical sensing, luminescence, and enhanced catalysis, etc. Among recent applications, the improvement of the localized field shows a vital role. The local field improvement significantly improves the far-field radiation of the weak signal and the optical detection beyond the diffraction limit. This local field relies on the size of the NPs.

Consequently, the plasmonic tweezer can manipulate the nanoscale particles twice as small as the traditional optical transmittance. A weak Raman scattering signal can be augmented by 103-fold or 102-fold. The NPs can proficiently catalyze the photochemical reaction because of the selective frequency absorption. Because of the high energy-transfer efficiency of the NPs, more than a 102-fold augmentation has been attained for the UCNP emissions. Through specific nanofabrication, researchers have also established plasmonic devices with an extraordinary optical recital. The chiral devices with ultra-high-resolution have shown outstanding optical response in high-frequency bands. Advanced nanofabrication methods are of prodigious significance in DNA-related studies and biological macromolecules. For example, focused ion beam lithography, direct laser writing, electron beam lithography, and glancing-angle deposition. The probable usage of plasmonic devices is as omnipresent as optics. These applications comprise novel methods to enhancements in sophisticated optical methods and unresolved matters. To a novel potential application in the future, fundamental research will continuously lead. The precise and accurate control of the functionalization of NPs, shape, and size are the current challenges for applications. Specifically, in the energy field, the keys to promoting a wide range of plasmonic technology applications are the excellent control of the dispersion of low-cost, large-scale synthesis of homogeneously shaped and sized NPs. A blooming and exciting era of plasmonics starts with real-life applications and more interesting scientific challenges soon with the parallel development of nanofabrication techniques besides the continuous theoretical advancement of plasmonics.

REFERENCES

1. Fernández-Domínguez, Antonio I., Francisco J. García-Vidal, and Luis Martín-Moreno. "Unrelenting plasmons." *Nature Photonics* 11, no. 1 (2017): 8–10. doi:10.1038/nphoton.2016.258.

2. Edel, Joshua B., Alexei A. Kornyshev, Anthony R. Kucernak, and Michael Urbakh. "Fundamentals and applications of self-assembled plasmonic nanoparticles at interfaces." *Chemical Society Reviews* 45, no. 6 (2016): 1581–1596. doi:10.1039/C5CS00576K.

3. Luo, Xiangang, and Lianshan Yan. "Surface plasmon polaritons and its applications." *IEEE Photonics Journal* 4, no. 2 (2012): 590–595. doi:10.1109/JPHOT.2012.2189436.

4. Chenthamara, Dhrisya, Sadhasivam Subramaniam, Sankar Ganesh Ramakrishnan, Swaminathan Krishnaswamy, Musthafa Mohamed Essa, Feng-Huei Lin, and M. Walid Qoronfleh. "Therapeutic efficacy of nanoparticles and routes of administration." *Biomaterials Research* 23, no. 1 (2019): 1–29. doi:10.1186/s40824-019-0166-x.

5. Pereira, Rui MS, Joel Borges, Georgui V. Smirnov, Filipe Vaz, and Mikhail I. Vasilevskiy. "Surface plasmon resonance in a metallic nanoparticle embedded in a semiconductor matrix: Exciton–plasmon coupling." *ACS Photonics* 6, no. 1 (2018): 204–210. doi:10.1021/acsphotonics.8b01430.

6. Homola, Jiří, and Marek Piliarik. "Surface plasmon resonance (SPR) sensors." In *Surface Plasmon Resonance Based Sensors*, pp. 45–67. Springer, Berlin, Heidelberg, 2006. doi:10.1007/5346_014.

7. Escobedo, Carlos, Alexandre G. Brolo, Reuven Gordon, and David Sinton. "Optofluidic concentration: Plasmonic nanostructure as concentrator and sensor." *Nano Letters* 12, no. 3 (2012): 1592–1596. doi:10.1021/nl204504s.

8. Wijaya, Edy, Cédric Lenaerts, Sophie Maricot, Juriy Hastanin, Serge Habraken, Jean-Pierre Vilcot, Rabah Boukherroub, and Sabine Szunerits. "Surface plasmon resonance-based biosensors: From the development of different SPR structures to novel surface functionalization strategies." *Current Opinion in Solid State and Materials Science* 15, no. 5 (2011): 208–224. doi:10.1016/j.cossms.2011.05.001.

9. Tang, Longhua, and Jinghong Li. "Plasmon-based colorimetric nanosensors for ultrasensitive molecular diagnostics." *ACS Sensors* 2, no. 7 (2017): 857–875. doi:10.1021/acssensors.7b00282.

10. Lesuffleur, Antoine, Hyungsoon Im, Nathan C. Lindquist, Kwan Seop Lim, and Sang-Hyun Oh. "Laser-illuminated nanohole arrays for multiplex plasmonic micro-array sensing." *Optics Express* 16, no. 1 (2008): 219–224. doi:10.1364/OE.16.000219.

11. Kelly, K. Lance, Eduardo Coronado, Lin Lin Zhao, and George C. Schatz. "The optical properties of metal nanoparticles: The influence of size, shape, and dielectric environment." (2003): 668–677. doi:10.1021/jp026731y.

12. Kreibig, Uwe, and Michael Vollmer. *Optical Properties of Metal Clusters.* Vol. 25. Springer Science & Business Media, 2013. doi:10.1007/978-3-662-09109-8.

13. Knight, Mark W., Nicholas S. King, Lifei Liu, Henry O. Everitt, Peter Nordlander, and Naomi J. Halas. "Aluminum for plasmonics." *ACS Nano* 8, no. 1 (2014): 834–840. doi:10.1021/nn405495q.

14. Ross, Michael B., and George C. Schatz. "Aluminum and indium plasmonic nano-antennas in the ultraviolet." *The Journal of Physical Chemistry C* 118, no. 23 (2014): 12506–12514. doi:10.1021/jp503323u.

15. Yang, Ankun, Alexander J. Hryn, Marc R. Bourgeois, Won-Kyu Lee, Jingtian Hu, George C. Schatz, and Teri W. Odom. "Programmable and reversible plasmon mode engineering." *Proceedings of the National Academy of Sciences* 113, no. 50 (2016): 14201–14206. doi:10.1073/pnas.1615281113.

16. Ritchie, Rufus H., E. T. Arakawa, J. J. Cowan, and R. N. Hamm. "Surface-plasmon resonance effect in grating diffraction." *Physical Review Letters* 21, no. 22 (1968): 1530. doi:10.1103/PhysRevLett.21.1530.

17. Otto, Andreas. "Excitation of nonradiative surface plasma waves in silver by the method of frustrated total reflection." *Zeitschrift für Physik A Hadrons and Nuclei* 216, no. 4 (1968): 398–410. doi:10.1007/BF01391532.

18. Barnes, William L., W. Andrew Murray, J. Dintinger, E. Devaux, and T. W. Ebbesen. "Surface plasmon polaritons and their role in the enhanced transmission of light through periodic arrays of subwavelength holes in a metal film." *Physical Review Letters* 92, no. 10 (2004): 107401. doi:10.1103/PhysRevLett.92.107401.

19. Singh, Ashok K. *Engineered Nanoparticles: Structure, Properties and Mechanisms of Toxicity.* Academic Press, 1st Edition, 2015, 1–554. doi:10.1016/B978-0-12-801406-6.00001-7.

20. Park, Jongnam, Jin Joo, Soon Gu Kwon, Youngjin Jang, and Taeghwan Hyeon. "Synthesis of monodisperse spherical nanocrystals." *Angewandte Chemie International Edition* 46, no. 25 (2007): 4630–4660. doi:10.1002/anie.200603148.

21. Pitarke, José María, Viatcheslav M. Silkin, Eugene V. Chulkov, and Pedro M. Echenique. "Surface plasmons in metallic structures." *Journal of Optics A: Pure and Applied Optics* 7, no. 2 (2005): S73. doi:10.1088/1464-4258/7/2/010.

22. Krajczewski, Jan, Karol Kołątaj, and Andrzej Kudelski. "Plasmonic nanoparticles in chemical analysis." *RSC Advances* 7, no. 28 (2017): 17559–17576. doi:10.1039/C7RA01034F.

23. Huang, Xiaohua, and Mostafa A. El-Sayed. "Gold nanoparticles: Optical properties and implementations in cancer diagnosis and photothermal therapy." *Journal of Advanced Research* 1, no. 1 (2010): 13–28. doi:10.1016/j.jare.2010.02.002.

24. Motl, N. E., A. F. Smith, C. J. DeSantis, and S. E. Skrabalak. "Engineering plasmonic metal colloids through composition and structural design." *Chemical Society Reviews* 43, no. 11 (2014): 3823–3834. doi:10.1039/C3CS60347D.

25. Chen, Shouhui, Dingbin Liu, Zhihua Wang, Xiaolian Sun, Daxiang Cui, and Xiaoyuan Chen. "Picomolar detection of mercuric ions by means of gold–silver core–shell nanorods." *Nanoscale* 5, no. 15 (2013): 6731–6735. doi:10.1039/C3NR01603J.

26. Huang, Haowen, Shenna Chen, Fang Liu, Qian Zhao, Bo Liao, Shoujun Yi, and Yunlong Zeng. "Multiplex plasmonic sensor for detection of different metal ions based on a single type of gold nanorod." *Analytical Chemistry* 85, no. 4 (2013): 2312–2319. doi:10.1021/ac303305j.

27. Lin, Qi, Pei Chen, Juan Liu, Yong-Peng Fu, You-Ming Zhang, and Tai-Bao Wei. "Colorimetric chemosensor and test kit for detection copper (II) cations in aqueous solution with specific selectivity and high sensitivity." *Dyes and Pigments* 98, no. 1 (2013): 100–105. doi:10.1016/j.dyepig.2013.01.024.

28. Sun, Yugang, and Younan Xia. "Gold and silver nanoparticles: A class of chromophores with colors tunable in the range from 400 to 750 nm." *Analyst* 128, no. 6 (2003): 686–691. doi:10.1039/B212437H.

29. Tong, Wenming, Michael J. Walsh, Paul Mulvaney, Joanne Etheridge, and Alison M. Funston. "Control of symmetry breaking size and aspect ratio in gold nanorods: Underlying role of silver nitrate." *The Journal of Physical Chemistry C* 121, no. 6 (2017): 3549–3559. doi:10.1021/acs.jpcc.6b10343.

30. Ghosh, Sujit Kumar, and Tarasankar Pal. "Interparticle coupling effect on the surface plasmon resonance of gold nanoparticles: From theory to applications." *Chemical Reviews* 107, no. 11 (2007): 4797–4862. doi:10.1021/cr0680282.

31. Wang, Ai-Jun, Han Guo, Ming Zhang, Dan-Ling Zhou, Rui-Zhi Wang, and Jiu-Ju Feng. "Sensitive and selective colorimetric detection of cadmium (II) using gold nanoparticles modified with 4-amino-3-hydrazino-5-mercapto-1, 2, 4-triazole." *Microchimica Acta* 180, no. 11–12 (2013): 1051–1057. doi:10.1007/s00604-013-1030-7.

32. Fratoddi, Ilaria. "Hydrophobic and hydrophilic Au and Ag nanoparticles. Breakthroughs and perspectives." *Nanomaterials* 8, no. 1 (2018): 11. doi:10.3390/nano8010011.

33. Requejo-Isidro, José, R. Del Coso, J. Solis, J. Gonzalo, and Carmen N. Afonso. "Role of surface-to-volume ratio of metal nanoparticles in optical properties of Cu: Al₂O₃ nanocomposite films." *Applied Physics Letters* 86, no. 19 (2005): 193104. doi:10.1063/1.1923198.

34. Bertolini, J-C., and J-L. Rousset. "Reactivity of Metal Nanoparticles." In *Nanomaterials and Nanochemistry*, pp. 281–304. Springer, Berlin, Heidelberg, 2008. doi:10.1007/978-3-540-72993-8_10.

35. Scholl, Jonathan A., Ai Leen Koh, and Jennifer A. Dionne. "Quantum plasmon resonances of individual metallic nanoparticles." *Nature* 483, no. 7390 (2012): 421–427. doi:10.1038/nature10904.

36. Nanda, K. K. "Size-dependent melting of nanoparticles: Hundred years of thermodynamic model." *Pramana* 72, no. 4 (2009): 617–628. doi:10.1007/s12043-009-0055-2.

37. Nanda, K. K., S. N. Sahu, and S. N. Behera. "Liquid-drop model for the size-dependent melting of low-dimensional systems." *Physical Review A* 66, no. 1 (2002): 013208. doi:10.1103/PhysRevA.66.013208.

38. Liu, Ying, and Quan Cheng. "Detection of membrane-binding proteins by surface plasmon resonance with an all-aqueous amplification scheme." *Analytical Chemistry* 84, no. 7 (2012): 3179–3186. doi:10.1021/ac203142n.

39. Špringer, Tomáš, Marek Piliarik, and Jiří Homola. "Real-time monitoring of biomolecular interactions in blood plasma using a surface plasmon resonance biosensor." *Analytical and Bioanalytical Chemistry* 398, no. 5 (2010): 1955–1961. doi:10.1007/s00216-010-4159-9.

40. Aslan, Kadir, Patrick Holley, Lydia Davies, Joseph R. Lakowicz, and Chris D. Geddes. "Angular-ratiometric plasmon-resonance based light scattering for bioaffinity sensing." *Journal of the American Chemical Society* 127, no. 34 (2005): 12115–12121. doi:10.1021/ja052739k.

41. Inci, Fatih, Onur Tokel, ShuQi Wang, Umut Atakan Gurkan, Savas Tasoglu, Daniel R. Kuritzkes, and Utkan Demirci. "Nanoplasmonic quantitative detection of intact viruses from unprocessed whole blood." *ACS Nano* 7, no. 6 (2013): 4733–4745. doi:10.1021/nn3036232.

42. Uludag, Yildiz, and Ibtisam E. Tothill. "Cancer biomarker detection in serum samples using surface plasmon resonance and quartz crystal microbalance sensors with nanoparticle signal amplification." *Analytical Chemistry* 84, no. 14 (2012): 5898–5904. doi:10.1021/ac300278p.

43. Bolduc, Olivier R., and Jean-Francois Masson. "Advances in surface plasmon resonance sensing with nanoparticles and thin films: Nanomaterials, surface chemistry, and hybrid plasmonic techniques." (2011): 8057–8062. doi:10.1021/ac2012976.

44. Oates, T. W. H., Herbert Wormeester, and Hans Arwin. "Characterization of plasmonic effects in thin films and metamaterials using spectroscopic ellipsometry." *Progress in Surface Science* 86, no. 11–12 (2011): 328–376. doi:10.1016/j.progsurf.2011.08.004.

45. Haes, Amanda J., Shengli Zou, Jing Zhao, George C. Schatz, and Richard P. Van Duyne. "Localized surface plasmon resonance spectroscopy near molecular resonances." *Journal of the American Chemical Society* 128, no. 33 (2006): 10905–10914. doi:10.1021/ja063575q.

46. Zhao, Jing, Lasse Jensen, Jiha Sung, Shengli Zou, George C. Schatz, and Richard P. Van Duyne. "Interaction of plasmon and molecular resonances for rhodamine 6G adsorbed on silver nanoparticles." *Journal of the American Chemical Society* 129, no. 24 (2007): 7647–7656. doi:10.1021/ja0707106.

47. Law, Wing-Cheung, Ken-Tye Yong, Alexander Baev, and Paras N. Prasad. "Sensitivity improved surface plasmon resonance biosensor for cancer biomarker detection based on plasmonic enhancement." *ACS Nano* 5, no. 6 (2011): 4858–4864. doi:10.1021/nn2009485.

48. Li, Jiangtian, Scott K. Cushing, Peng Zheng, Fanke Meng, Deryn Chu, and Nianqiang Wu. "Plasmon-induced photonic and energy-transfer enhancement of solar water splitting by a hematite nanorod array." *Nature Communications* 4, no. 1 (2013): 1–8. doi:10.1038/ncomms3651.

49. Ebbesen, Thomas W., Henri J. Lezec, H. F. Ghaemi, Tineke Thio, and Peter A. Wolff. "Extraordinary optical transmission through sub-wavelength hole arrays." *Nature* 391, no. 6668 (1998): 667–669. doi:10.1038/35570.

50. Brolo, Alexandre G., Reuven Gordon, Brian Leathem, and Karen L. Kavanagh. "Surface plasmon sensor based on the enhanced light transmission through arrays of nanoholes in gold films." *Langmuir* 20, no. 12 (2004): 4813–4815. doi:10.1021/la0493621.

51. Liu, Na, Thomas Weiss, Martin Mesch, Lutz Langguth, Ulrike Eigenthaler, Michael Hirscher, Carsten Sonnichsen, and Harald Giessen. "Planar metamaterial analogue of electromagnetically induced transparency for plasmonic sensing." *Nano Letters* 10, no. 4 (2010): 1103–1107. doi:10.1021/nl902621d.

52. Lesuffleur, Antoine, Hyungsoon Im, Nathan C. Lindquist, Kwan Seop Lim, and Sang-Hyun Oh. "Laser-illuminated nanohole arrays for multiplex plasmonic micro-array sensing." *Optics Express* 16, no. 1 (2008): 219–224. doi:10.1364/OE.16.000219.

53. Im, Hyungsoon, Jamie N. Sutherland, Jennifer A. Maynard, and Sang-Hyun Oh. "Nanohole-based surface plasmon resonance instruments with improved spectral resolution quantify a broad range of antibody-ligand binding kinetics." *Analytical Chemistry* 84, no. 4 (2012): 1941–1947. doi:10.1021/ac300070t.

54. De Leebeeck, Angela, L. K. Swaroop Kumar, Victoria De Lange, David Sinton, Reuven Gordon, and Alexandre G. Brolo. "On-chip surface-based detection with nanohole arrays." *Analytical Chemistry* 79, no. 11 (2007): 4094–4100. doi:10.1021/ac070001a.

55. Pang, Lin, Grace M. Hwang, Boris Slutsky, and Yeshaiahu Fainman. "Spectral sensitivity of two-dimensional nanohole array surface plasmon polariton resonance sensor." *Applied Physics Letters* 91, no. 12 (2007): 123112. doi:10.1063/1.2789181.

56. Liu, Zhong De, Yuan Fang Li, Jian Ling, and Cheng Zhi Huang. "A localized surface plasmon resonance light-scattering assay of mercury (II) on the basis of Hg^{2+}– DNA complex induced aggregation of gold nanoparticles." *Environmental Science & Technology* 43, no. 13 (2009): 5022–5027. doi:10.1021/es9001983.

57. Gao, Chuanbo, Zhenda Lu, Ying Liu, Qiao Zhang, Miaofang Chi, Quan Cheng, and Yadong Yin. "Highly stable silver nanoplates for surface plasmon resonance biosensing." *Angewandte Chemie International Edition* 51, no. 23 (2012): 5629–5633. doi:10.1002/anie.201108971.

58. Lahav, Michal, Alexander Vaskevich, and Israel Rubinstein. "Biological sensing using transmission surface plasmon resonance spectroscopy." *Langmuir* 20, no. 18 (2004): 7365–7367. doi:10.1021/la0489054.

59. Méjard, Régis, Jakub Dostálek, Chun-Jen Huang, Hans Griesser, and Benjamin Thierry. "Tuneable and robust long-range surface plasmon resonance for biosensing applications." *Optical Materials* 35, no. 12 (2013): 2507–2513. doi:10.1016/j.optmat.2013.07.011.

60. Vala, M., S. Etheridge, J. A. Roach, and J. Homola. "Long-range surface plasmons for sensitive detection of bacterial analytes." *Sensors and Actuators B: Chemical* 139, no. 1 (2009): 59–63. doi:10.1016/j.snb.2008.08.029.

61. Sarid, Dror. "Long-range surface-plasma waves on very thin metal films." *Physical Review Letters* 47, no. 26 (1981): 1927. doi:10.1103/PhysRevLett.47.1927.

62. Li, Ming, Scott K. Cushing, Jianming Zhang, Jessica Lankford, Zoraida P. Aguilar, Dongling Ma, and Nianqiang Wu. "Shape-dependent surface-enhanced Raman scattering in gold–Raman-probe–silica sandwiched nanoparticles for biocompatible applications." *Nanotechnology* 23, no. 11 (2012): 115501. doi:10.1088/0957-4484/23/11/115501.

63. Zhang, Xiao-Bing, Rong-Mei Kong, and Yi Lu. "Metal ion sensors based on DNAzymes and related DNA molecules." *Annual Review of Analytical Chemistry* 4 (2011): 105–128. doi:10.1146/annurev.anchem.111808.073617.

64. Kim, Youngjin, Robert C. Johnson, and Joseph T. Hupp. "Gold nanoparticle-based sensing of "spectroscopically silent" heavy metal ions." *Nano Letters* 1, no. 4 (2001): 165–167. doi:10.1021/nl0100116.

65. Chai, Fang, Chungang Wang, Tingting Wang, Lu Li, and Zhongmin Su. "Colorimetric detection of Pb^{2+} using glutathione functionalized gold nanoparticles." *ACS Applied Materials & Interfaces* 2, no. 5 (2010): 1466–1470. doi:10.1021/am100107k.

66. Xue, Xuejia, Feng Wang, and Xiaogang Liu. "One-step, room temperature, colorimetric detection of mercury (Hg^{2+}) using DNA/nanoparticle conjugates." *Journal of the American Chemical Society* 130, no. 11 (2008): 3244–3245. doi:10.1021/ja076716c.

67. Homola, Jiří. "Present and future of surface plasmon resonance biosensors." *Analytical and Bioanalytical Chemistry* 377, no. 3 (2003): 528–539. doi:10.1007/s00216-003-2101-0.

68. Liang, Hongyan, Haiguang Zhao, David Rossouw, Wenzhong Wang, Hongxing Xu, Gianluigi A. Botton, and Dongling Ma. "Silver nanorice structures: Oriented attachment-dominated growth, high environmental sensitivity, and real-space visualization of multipolar resonances." *Chemistry of Materials* 24, no. 12 (2012): 2339–2346. doi:10.1021/cm3006875.

69. Kong, Biao, Anwei Zhu, Yongping Luo, Yang Tian, Yanyan Yu, and Guoyue Shi. "Sensitive and selective colorimetric visualization of cerebral dopamine based on double molecular recognition." *Angewandte Chemie International Edition* 50, no. 8 (2011): 1837–1840. doi:10.1002/anie.201007071.

70. Elghanian, Robert, James J. Storhoff, Robert C. Mucic, Robert L. Letsinger, and Chad A. Mirkin. "Selective colorimetric detection of polynucleotides based on the distance-dependent optical properties of gold nanoparticles." *Science* 277, no. 5329 (1997): 1078–1081. doi:10.1126/science.277.5329.1078.

71. Kim, Youngjin, Robert C. Johnson, and Joseph T. Hupp. "Gold nanoparticle-based sensing of "spectroscopically silent" heavy metal ions." *Nano Letters* 1, no. 4 (2001): 165–167. doi:10.1021/nl0100116.

72. Kim, Youngjin, Robert C. Johnson, and Joseph T. Hupp. "Gold nanoparticle-based sensing of "spectroscopically silent" heavy metal ions." *Nano Letters* 1, no. 4 (2001): 165–167. doi:10.1021/nl0100116.

73. Giljohann, David A., Dwight S. Seferos, Weston L. Daniel, Matthew D. Massich, Pinal C. Patel, and Chad A. Mirkin. "Gold nanoparticles for biology and medicine." *Angewandte Chemie International Edition* 49, no. 19 (2010): 3280–3294. doi:10.1002/anie.200904359.

74. Lee, Jae-Seung, Min Su Han, and Chad A. Mirkin. "Colorimetric detection of mercuric ion (Hg^{2+}) in aqueous media using DNA-functionalized gold nanoparticles." *Angewandte Chemie International Edition* 46, no. 22 (2007): 4093–4096. doi:10.1002/anie.200700269.

75. Xue, Xuejia, Feng Wang, and Xiaogang Liu. "One-step, room temperature, colorimetric detection of mercury (Hg^{2+}) using DNA/nanoparticle conjugates." *Journal of the American Chemical Society* 130, no. 11 (2008): 3244–3245. doi:10.1021/ja076716c.

76. Xu, Xiaoyang, Weston L. Daniel, Wei Wei, and Chad A. Mirkin. "Colorimetric Cu^{2+} detection using DNA-modified gold-nanoparticle aggregates as probes and click chemistry." *Small* 6, no. 5 (2010): 623–626. doi:10.1002/smll.200901691.

77. Sun, Yugang, and Younan Xia. "Increased sensitivity of surface plasmon resonance of gold nanoshells compared to that of gold solid colloids in response to environmental changes." *Analytical Chemistry* 74, no. 20 (2002): 5297–5305. doi:10.1021/ac0258352.

78. Yguerabide, Juan, and Evangelina E. Yguerabide. "Light-scattering submicroscopic particles as highly fluorescent analogs and their use as tracer labels in clinical and biological applications: II. Experimental characterization." *Analytical Biochemistry* 262, no. 2 (1998): 157–176. doi:10.1006/abio.1998.2760.

3 Biogenic Silver and Gold Nanostructures as SPR Based Sensors for the Detection of Toxic Metal Ions in Aqueous Media

3.1 INTRODUCTION

With diverse localities, composition, and various concentrations on the earth's crust, the accessibility of heavy metals varies adjacent, which is natural. The heavy metal ions present in aqueous systems sources graves hazards to the environment and human health. The rapid increase of heavy metal ions arises because of industrial and mining activities' progress. The natural sources can carefully increase the discharge of toxic metals into waters, for example, soil, rainwater, erosion from rock, and weathering discharges. For example, lead, cobalt, arsenic, copper, cadmium, magnesium, nickel, and mercury can cause pollution in soil, air, and water, consistent with severe toxic effects and health issues [1–5]. The toxicants of heavy metal ions can occur as complexes, elements, or ions and contrast to water density, and the heavy metal ions have five times greater density up to $4\,g\,cm^{-3}$ [6]. Several conventional analytical methods have been used at very low- and ultra-low-level detection of heavy metals, such as inductively coupled plasma optical emission spectroscopy (ICP-OES), UV–vis spectroscopy, capillary electrophoresis (CE), liquid chromatography, X-ray fluorescence spectroscopy (XFS), anodic stripping voltammetry (ASV), atomic absorption spectroscopy (AAS), inductively coupled plasma mass spectroscopy (ICP-MS), and microprobes (MP). Several limitations suffer from these analytical methods, such as preparation of the sample, pre-concentration processes, cleanup processes, professional personnel, and expensive instruments [7–10]. As an alternative for detecting metal ions, electrochemical, fluorescence, and colorimetric detection are the approaches that are facile, inexpensive, and time-saving techniques [11–13]. In detecting heavy metals, nanomaterial-based sensors have shown great potential due to their large surface area, high surface reactivity, high catalytic efficiency, and strong adsorption capacity. Various nanomaterials have been used to synthesize nanosensors, such

DOI: 10.1201/9781003128281-3

as carbon-based nanomaterials, MNPs, metal oxide NPs, polymeric nanomaterials, and silicon nanomaterials. The nanosensor's enhanced sensitivity and selectivity are due to the high reactivity, large surface-to-volume ratio, high degree of functionalization, and size-dependent properties [14].

Nano components have been modified to improve selectivity and sensitivity through covalent functionality and anchoring of organic ligands. The nanomaterials included in chemical sensors have significantly enhanced their reproducibility and detection limit [10]. In this context, it extremely deliberated the electronic and optical properties of nanomaterials. This chapter focuses on using many nanomaterials, i.e., AuNPs and AgNPs, used as sensors using varied approaches to sense heavy metal ions.

3.2   HEAVY METAL ION DETECTION USING BIOSYNTHESIZED AgNPs

Earlier, owing to its ecofriendly, reliable, and sustainable nature, the synthesis of nanomaterials by the green route is fascinating extensive consideration. The synthesis of NPs thru the biological approach is pondered as a powerful method to decrease severe destructive effects escorted thru traditional synthesis methods for the synthesis of NPs used in laboratories and industries. The area of synthesis of NPs using the green route has been discovered for the utilization of essential phytochemicals, e.g., aldehydes, amides, flavonoids, alkaloids, ketones, phenols, ascorbic acids, and carboxylic acids, and some vital biological components, such as yeast, plants, fungi, and bacteria as reducing and capping agents [15–19]. Green synthesis approaches are contingent on many reaction parameters such as pH, temperature, and solvent system and primarily depend on natural circumstances. In most cases, to synthesize purely metal NPs or metal oxide, phytochemicals are vital species. Using lemon and orange extract were used, which includes citric acid and ascorbic acid (AA) as a reducing agent [20], Ravi et al. (2013) developed green and sustainable methods for reducing Ag and Au ions into simple MNPs. For reduction, both lemon and orange extract are robust contenders because of their reducing capabilities.

Nevertheless, for reducing Ag and Au ions into Au and AgNPs, lemon extract functioned under sunlight contact, whereas orange extract was the more potent reducing agent. These Au and AgNPs were further used for selective and sensitive sensing at a micromolar concentration of hazardous Hg^{2+} metal ions. Using the lemon fruit extract-based AgNPs, the detection of Hg^{2+} has been done at a 3.2–8.5 pH range.

In another study, Annadhasan et al. (2014) reported L-tyrosine stabilized Ag and Au NPs to detect Pb^{2+}, Hg^{2+}, and Mn^{2+} in an aqueous medium [21]. The morphology and shape of particles can affect to greater amount thru the controlled reaction parameters. However, –COO group in L-tyrosine was fixed to the apparent NPs surface sturdily, even if, to reduce metal ion into metal NPs, –COO⁻ and OH groups are the key tools. Inclusively, the sensing of Mn^{2+} and Hg^{2+} ions having 16 nM of the detection limit for both ions confirmed the sensitivity of AgNPs for the detection method. Besides, AuNPs showed sensitivity up to 16 and 53 nM of LOD to detect Pb^{2+} and Hg^{2+} ions, correspondingly.

For the colorimetric detection of Hg^{2+} ions, AgNPs stabilized by polyethylene glycol poly vinylpropyl (PEG-PVP) by Ahmed et al. (2014), which was further characterized under UV–vis spectroscopy by SPR using a green synthesis reduction approach [22]. The inverse and linear relationship was observed over the range of 10–1 ppm among absorbance intensity of the Hg^{2+} ions and AgNPs concentration. However, with 1 ppm LOD, the absorption was observed at 411 nm.

Likewise, for the electrochemical sensing of metals using anthocyanin capped carbon paste, synthesized AuNPs have been done [23]. Phytochemicals in anthocyanin worked as a capping agent extracted from black rice, whereas as a stabilizing agent, gum Arabic was utilized in the synthesis method. The results displayed that the anthocyanin capped carbon paste electrode (GNMACPE) changed AuNPs were found best suitable to detect Cu^{2+}, Pb^{2+}, and Cd^{2+}. The modified AuNPs were most suitable for the voltammetric determination of metal ions with different surfactants simultaneously with electroactive surface area.

Furthermore, using AgNPs functionalized with biopolymer xylan, Hg^{2+} ions were detected by Luo et al. (2015), which achieved soothing properties by stabilizing and reducing metal ions. Instead, microwave irradiation and Tollens reaction were used for the detection method [24]. The outcome displayed that the biologically developed AgNPs were mono-dispersed because of the package of xylan and below 20–35 nm [24]. In a mixed metal ion solution, the selective detection of Hg^{2+} ions was done. Though on the detection of Hg^{2+} ion, there was no remarkable effect of added metal ions, the LOD was measured as 4.6 nM for Hg^{2+} ion. Instead, for detecting Hg^{2+} ions, xylan-AgNPs were also tested for the real water samples as a sensing probe. In surface functionalization, poly-dopamine has been widely utilized due to different adhesive functions of mussels.

The green-synthesized Annadhasan has used AgNPs and Rajendiran (2015) to detect Hg^{2+} ions in tap water [25] selectively. N-cholyl-L-cysteine (NaCysC) as a capping and reducing agent has been used to synthesize AgNPs under ambient sunlight irradiation, without using other hazardous chemicals. The average size of AgNPs was 7.8 nm. Through the green-synthesized technique, the detection of Hg^{2+} ions on AgNPs was carried out with LOD of 8 nM at a low concentration through the green-synthesized procedure.

Burmese grape's fruit juice was used to detect Hg^{2+} ions under sonochemical conditions as a stabilizing agent for synthesizing AgNPs [26]. The results showed that the surface-adsorbed functionalities of AgNPs determined the modification in aggregation actions of these nanomaterials. In comparison to citrate, PVP, SDS, and hexadecyltrimethylammonium bromide (CTAB) stabilized AgNPs under sonochemical conditions ultimately upsurge the metal ion sensing capabilities of the NPs. The average size of mono-dispersed AgNPs was 8 nm, and at pH 3.73–11.18, the LOD was recorded 47.6 μM for Hg^{2+} ions. Correspondingly, N-cholyl-L-valine (NaValC)-reduced AuNPs have been developed to detect Co^{2+} and Ni^{2+} ions in water under natural sunlight irradiation [27]. The size and shape of developed AuNPs capped with Na-ValC were perceived through altering the concentration from low to high of the capping agent. The average size of AuNPs was 40 nm at a lower concentration, whereas the size of well-dispersed stabilized AuNPs was noticeably reduced to 8.0 nm at maximum concentration. The lipophilic and hydrophilic groups were utilized for the stability of the AuNPs. In contrast, the OH and NH–C=O groups were involved in Na-ValC to reduce metal ions. For Ni^{2+} and Cu^{2+} ions, the LOD and the linear regression coefficient (R^2) were 10 nM and 0.9870, 10 nM, and 0.9806.

AgNPs functionalized with mussel-inspired protein have been utilized for colorimetric detection of Cu^{2+} and Pb^{2+} ions. A self-polymerized 3,4-dihydroxy L-phenyl-alanine (poly DOPA) was utilized to prepare AgNPs as reducing and stabilizing agents. The size of poly DOPA-AgNPs was 25 nm, and these were well dispersed because of the poly DOPA. The color change of developed NPs solutions from yellow to brown is due to the aggregation of poly DOPA-AgNPs, further confirmed by the transmission electron microscopy (TEM) images. The LOD was found 9.4×10^{-5} and 8.1×10^{-5} M for Pb^{2+} and Cu^{2+} ions.

AgNPs were synthesized using sugar bodies to detect the Hg^{2+} and Fe^{2+} metal ions calorimetrically, i.e., dextrose and maltose completely green method [28]. The size of sugar-reduced gelatin-capped AgNPs was found to be 18.17 ± 1 and 11.88 ± 1 nm, correspondingly. Usually, in detecting Hg^{2+} and Fe^{2+} metal ions, dextrose-reduced sugar was more sensitive than maltose at maximum metal ion concentration. Toward Hg^{2+} and Fe^{2+} ions, the LOD for maltose reduced AgNPs was found 0.1 M. In contrast, dextrose-reduced AgNPs it selectively detected at 0.001 M. For the detection of Hg^{2+} ions, at a linear range of 10^{-1} to 10^{-5} M, the sugar-reduced AgNPs sensors showed good sensitivity up to 10^{-12} M.

For the rapid and selective detection of Hg^{2+} ions in aqueous solutions, egg white-capped AgNPs have been developed via a green method [29]. Egg white exhibited a high ability to develop AgNPs efficiently and improved the fixed binding positions to Hg^{2+} ions associated with permissible thiol groups. The egg white-capped AgNPs can selectively detect the Hg^{2+} metal ions in the presence of other metal ions. Instead, the LOD up to 300 nM showed by the sensing probe in a dynamic range from 1,000 to 50,000 nM.

Under sunlight irradiation, the detection of Hg^{2+} ions has been done by Firdaus et al. (2017), developing fruit extract of Carica papaya stabilized AgNPs [30]. Similarly, for reducing the metal ions into their zero-oxidation state, phenolic and ascorbic acids are utilized. Bioactive compounds soluble in the water acted as bio-reductants to synthesize AgNPs from papaya fruit by reducing Ag^+. In various metals, the detection probe has checked the selectivity studies of Hg^{2+} ions. Through the color change from yellowish-brown to a colorless solution, AgNPs exhibited high selectivity toward Hg^{2+} ions by the colorimetric method. These color changes were observed because of Ag metal to Ag^+ oxidation for the synthesis of AgNPs. Table 3.1 displays the details of used NPs synthesized biologically with various recognition methods to detect heavy metals.

Extracts of citrus fruit (citrus limetta lemon (Cl-2) and sweet orange citrus lemon (Cl-1)) have been used to synthesize crystalline AgNPs and AuNPs.

TABLE 3.1

Biologically Synthesized Nanomaterials Used to Detect Heavy Metals

Nanomaterials	Size	Detection Method	Metals	LOD	References
PEG-PVP stabilized (AgNPs)	-	Colorimetry	Hg^{2+}	1 ppm	[22]
sodium salt of N-cholyl-L-cysteine (NaCysC) (AgNPs)	7.8 nm	Colorimetry	Hg^{2+}	8.0 nM	[25]
AuNPs/CNFs	5–15 nm	Electrochemical	Cd^{2+} Pb^{2+} Cu^{2+}	0.1 μM	[31]
Sugar-reduced gelatin-capped (AgNPs)	11.88±1 and 18.17±1 nm	Colorimetry	Hg^{2+} Fe^{2+}	10–12 M	[28]
Biopolymer xylan (AgNPs)	20–35 nm	Colorimetry	Hg^{2+}	4.6 nM	[24]
Egg white-capped (AgNPs)	3.8±0.5 nm	Colorimetry	Hg^{2+}	300 nM	[29]
Citrus fruit extract (AuNPs) and (AgNPs)	10–25 nm 20–35 nm	Colorimetry	Hg^{2+}	-	[20]
Anthocyanin-based (AuNPs)	31–52 nm	Voltammetry	Pb^{2+} Cd^{2+} Cu^{2+}	9.178, 86.327, 85.373 μg L^{-1}	[23]
N-cholyl-L-valine (NaValC) reduced gold (AuNPs)	40 nm	Colorimetry	Ni^{2+} Co^{2+}	10 nM	[27]
Protein based (AgNPs)	25 nm	Colorimetry	Pb^{2+} Cu^{2+}	9.4×10^{-5} and 8.1×10^{-5} M	[32]
Carica papaya fruit extract (AgNPs)	35–50 nm	Colorimetry	Hg^{2+}	-	[30]
L-tyrosine (AuNPs) and (AgNPs)	32±2 nm 36±2 nm	Colorimetry	Hg^{2+} Pb^{2+} Mn^{2+}	53 nM 16 nM	[21]
Burmese grape (AgNPs)	5–10 nm	Colorimetry	Hg^{2+}	47.6 μM	[26]

The synthesized NPs were used to detect hazardous aqueous metal ions. Cl-2 and Cl-1 are mostly contained ascorbic and citric acid and displayed diverse reducing capabilities to Ag and Au ions into their respective NPs. Cl-1 can only reduce Au ions into AuNPs only in the absence of sunlight. In contrast, the existence of sunlight is mandatory to reduce Ag ions into AgNPs. Apart from this, Cl-2 can reduce both Ag and Au ions into AgNPs and AuNPs, correspondingly in the absence of sunlight. The biologically fabricated AgNPs were used for colorimetric detection of Hg^{2+} ions at micromolar concentrations. Significantly, to detect Hg^{2+} ions using a biologically developed Cl-1-AgNPs sensor probe, the pH range of 3.2 and 8.5 is the optimum pH. The synthesized Cl-1-AgNPs sensor displayed excellent antibacterial properties [33]. Figure 3.1 demonstrates the schematic representation of the detection mechanism using green-synthesized AgNPs and AuNPs.

The usage of green-synthesized AgNPs has also been reported to detect Zn^{2+} ions by a cost-effective, selective, and simple colorimetric sensor. For the synthesis of AgNPs through the room temperature method and heat treatment method, the leaf extract of Amomum subulatum has been used. The AgNPs synthesized by the heat treatment method showed effective outcomes. Various analytical tools have been used to characterize as-synthesized AgNPs, such as Fourier transform infrared spectroscopy (FTIR) analysis, SEM, and UV–vis spectroscopy showing a sharp absorption band at 425 nm. For Zn^{2+} metal ion sensing, the developed AgNPs showed a suitable colorimetric sensing property, which was confirmed by

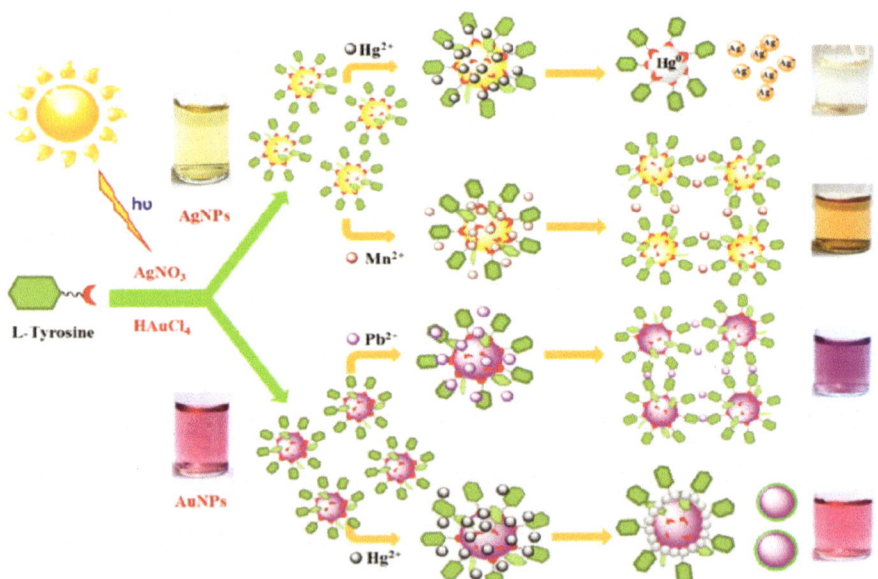

FIGURE 3.1 Detection of heavy metal ions using green-synthesized Ag and AuNPs. (Reprinted with permission from Ref. [21]. Copyright 2014 American Chemical Society.)

a reduced absorption intensity associated with altering the color of NPs from yellowish-brown to colorless solution. In the concentrations ranges from 1×10^{-5} to 8×10^{-5} M, the developed probe showed outstanding linear response for Zn^{2+} ions with 0.996 value of regression coefficient (R^2). The detection method showed the 3.5×10^{-6} M of LOD for Zn^{2+} ions. In a mixture of heavy metal ions, the synthesized sensor selectively detects Zn^{2+} ions. To sense Zn^{2+} ions, it effectively applied the synthesized sensor to drinking water samples. Due to the green method's cost-effectiveness, the as-synthesized AgNPs are beneficial for deployed areas. The synthesized AgNPs could do efficiently from the contaminated water the removal of Zn^{2+} ions [34].

By synthesizing green-fabricated AgNPs in a similar study, the selective sensing of Hg^{2+} ions has been done using LSPRs. An onion extract was utilized to stabilize and reduce the agent for the fabrication process, forming photoinduced green crystalline AgNPs. Through the presence of bioligands as stabilizing and reducing agents in the presence of ultrasound irradiation, the synthesis of NPs is enhanced. For the characterization of NPs, various analysis methods are used efficiently, e.g., UV–vis, TEM, XRD, FTIR, and SEM/EDAX. The TEM analysis confirmed the spherical nature of green-fabricated AgNPs. The SEM analysis showed uniformity, high-density, and spherical AgNPs. The XRD patterns of AgNPs showed their face-centered cubic arrangement. The EDAX analysis confirmed the development of impurity-free AgNPs. The detection of Hg^{2+} ions has been done on green-synthesized AgNPs by using a phosphate buffer medium. It also experimented with the detection probe for the selectivity and sensitivity studies through the reduction of LSPRs. The absorbance changed abruptly, which confirms Ag^+ reduction to $Ag(0)$. A progressive reduction of green-synthesized AgNPs occurs by a minor upsurge in Hg^{2+} concentration. The as-synthesized green AgNPs were also used in real water samples to detect harmful Hg^{2+} ions colorimetrically. The outcomes displayed below 6% of good recovery and an RSD of over 92% [35].

Agaricus bisporus has been used as a stabilizing and reducing agent due to its ecofriendly nature for AgNPs (AgNP-AB) fabrication to sense heavy metals. For the optical and electrochemical sensing of Hg^{2+} ions, they effectively used the developed AgNPs. For the synthesis of AgNP-AB, the microwave reactor has been used. The Hg^{2+} ions have been detected on synthesized AgNP with no use of further modifiers or sophisticated instrumentation. A color change from brown to black was observed due to the formation of the AgNP-AB-Hg^{2+} complex by the aggregation of AgNP-AB.

Moreover, the detection ability of AgNP toward metal ions has been revealed thru the usage of electrochemical studies. A platinum electrode (AgNP-AB/PE) changed with AgNP-AB was synthesized to detect toxic Hg^{2+} ions rapidly. The detection probe showed 2.1×10^{-6} M of LOD toward Hg^{2+} ions. The as-synthesized AgNP-AB/PE sensor was also examined for sensitivity studies toward Hg^{2+} ions in a mixture of metal ions. The detection probe also checked for the real water samples. The synthesized AgNP-AB is a highly versatile and promising sensor for various biological applications [36]. After adding many concentrations of heavy metal ions for the selectivity studies, the change in SPR peaks of UV–vis spectra of AgNPs and AuNPs is shown in Figure 3.2. The plot of absorbance ratios of metal ion versus AuNPs concentrations is shown in the inset image.

Correspondingly, for detecting Cu^{2+} ion and Hg^{2+} ions, the green-synthesized AgNPs have been utilized as sensor probes using two methods. Because of the complex formation of the metal ion, the alteration in absorbance occurs,

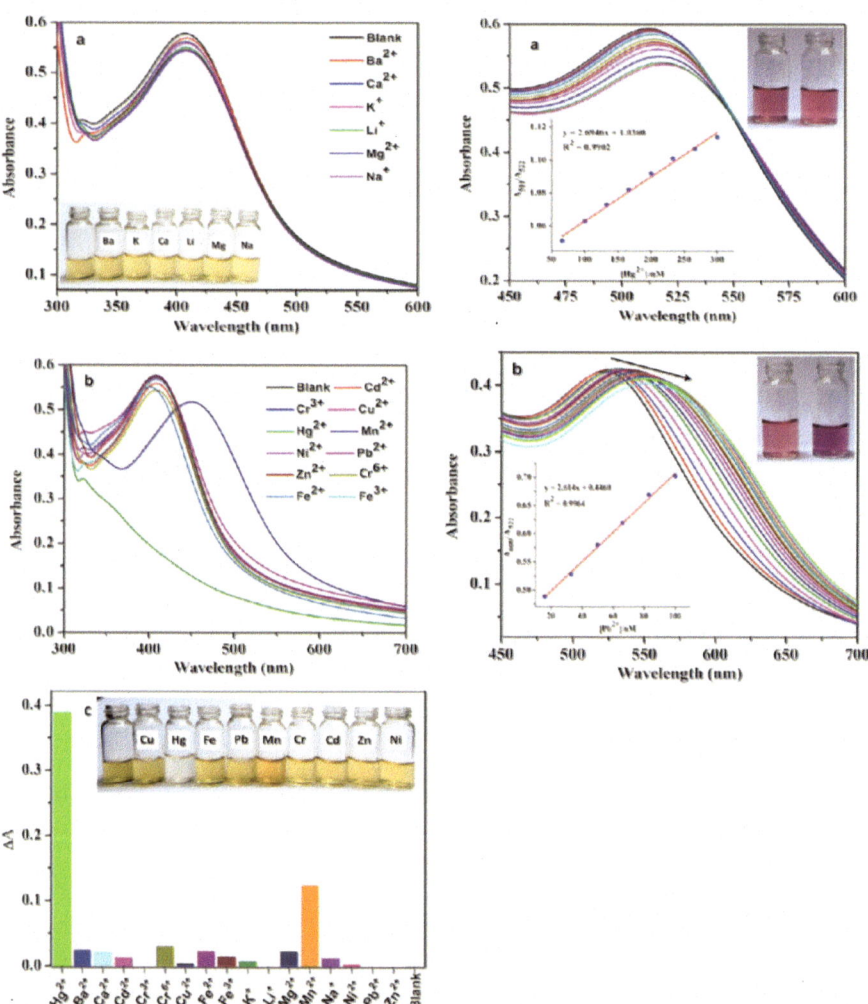

FIGURE 3.2 After adding various concentrations of heavy metal ions, the observed UV–visible spectra of AgNPs and AuNPs. The plot of absorbance ratios of metal ion concentrations versus AuNPs (Inset). (Reprinted with permission from Ref. [21]. Copyright 2014 American Chemical Society.)

demonstrating the Cu^{2+} ion detection. The linearity and sensitivity of the detection were examined by using numerous concentrations of Cu^{2+} ions. Another peak was detected at 770 nm and the peak at 406 nm due to the AgNP. In an aqueous solution, by increasing Cu^{2+} ion concentration, the absorbance at 406 nm peak condensed and increased progressively up to 770 nm. The absorption coefficients ratio of the concentration of Cu^{2+} ions versus these two peaks (Ex770/406) in 1 M concentration of water for the quantitative assessment of the amount of Cu^{2+} ions via a calibration curve. 3-mercapto-1, 2-propanediol (MPD) have been utilized to functionalize the developed AgNP to detect Hg^{2+} ions in aqueous media through the colorimetric method. They observed a new peak at around 606 nm after adding Hg^{2+} ion in AgNP solution functionalized with MPD (MPD-AgNP), accompanied by the peak at 404 nm. The detection of Hg^{2+} ion was confirmed due to the aggregations of MPD-AgNP through dipropionate ion by new peak appearance. The quantitative estimation of Hg^{2+} present in 1 M level of water was confirmed through plotting the calibration curve among the concentration of Hg^{2+} ions and ratios of the absorption coefficients of these two peaks (Ex404/606) [37].

An ecofriendly and cost-effective method was used to develop the robust AgNPs using an aqueous extract of *Murraya koenigii* (AEM) as a stabilizing and reducing agent. After 30 min of exposer of a mixer of the AEM reaction mixture and $AgNO_3$ solution to sunlight, the biosynthesis of AgNPs was detected by the appearance of dark brown color, which was further confirmed by a sharp SPR band at 430 nm in UV–vis spectroscopy. The optimum conditions for the biogenic AgNPs are 4.0 mM $AgNO_3$ concentration, 2.0% (v/v) of AEM inoculum dose, and 30 min contact time of sunlight exposure. The average size of developed AgNPs was found to be 8.6 nm with a spherical shape, which was confirmed by the TEM analysis. The face-centered cubic crystal lattice nature of metallic silver confirmed by Bragg's diffraction pattern appeared in the XRD pattern of AgNPs at (111), (200), (220), and (311), which also show their crystalline nature. The surface texture of synthesized AgNPs has an average roughness up to 1.8 nm, confirmed by the atomic force microscopy (AFM) analysis. In the synthesis of AgNPs, the participation of several functional groups was established by the recorded FTIR spectra between 4,000 and 400 cm^{-1}. The colorimetric detection of Hg^{2+} ions was confirmed with the linear relationship among SPR band intensity having 50 nm to 500 μM of the linear range of Hg^{2+} ions concentration by using as-synthesized AgNPs. An oxidation-reduction mechanism was also proposed among Hg^{2+} ions and AgNPs based on experimental outcomes [38].

Recently, the synthesis of cost-effective NPs as colorimetric sensors has fascinated widespread attention. The NPs can be exploited over a diverse range of targets in various environmental and biological analyses, e.g., heavy metal cations, anions, proteins, etc. Flower extract of *Neolamarckia cadamba* has been used to synthesize AgNPs in an aqueous medium to detect Hg^{2+} ions colorimetrically. Even at 0.000005–0.00004 M concentration of Hg^{2+} ions, the color of NPs changed after the addition of Hg^{2+} ions into the NPs solution quickly from yellowish-brown to colorless. In contrast to other ions, for example, Ni^{2+}, Pb^{2+},

Cd^{2+}, Cu^{2+}, Co^{2+}, Fe^{3+}, and Zn^{2+}, the biosynthesized AgNPs have more selectivity for Hg^{2+} ions, which was confirmed by the sensitivity and selectivity studies [39].

Because of the hazardous properties of heavy metals in human health and the ecosystem, a colorimetric method has been developed to sense heavy metals. Using roots extract of *Bistorta amplexicaulis*, functionalized AgNPs were synthesized to detect Hg^{2+} and Pb^{2+} ions. The as-synthesized biogenic AgNPs were well-characterized by using numerous analytical methods, such as UV–vis, XRD, FTIR, Zetasizer, and AFM. The photophysical potential of synthesized AgNPs was measured using colorimetric assay and absorption spectroscopy for common metal ions. The SPR band of AgNPs shifted toward a hypsochromic shift in the case of Hg^{2+} ions. The detection of Hg^{2+} ions was also confirmed thru color change from dark brown to light yellow by naked eyes. In the absorbance of AgNPs, a substantial reduction was recorded to mix Pb^{2+} ions to the AgNPs solution. The LOD for Hg^{2+} and Pb^{2+} ions, which display the detection probe's high sensitivity, were found to be 8.0×10^{-7} and 2.0×10^{-7} M on AgNPs-based colorimetric sensor, respectively. For the degradation of methyl orange dye, AgNPs exhibited promising catalytic activity. The outcomes divulge the potential applications of *Bistorta amplexicaulis* stabilized AgNPs as a catalyst for the degradation of methyl orange and an effective colorimetric sensor [40].

3.3 DETECTION HEAVY METAL IONS USING BIOSYNTHESIZED AuNPs

In aqueous solutions, the biogenic AuNPs could be detected as a twofold, practical, plasmon resonance, and colorimetric sensing technique to detect heavy metals. For the green development of AuNPs, the leaf extract of *Camellia sinensis* was used as a reducing agent. For example, various analytical tools, TEM, FTIR, and UV–vis, were used to characterize the green-synthesized AuNPs. A pink-reddish color was attained for the colloidal solution AuNPs, having SPRs peaks between 529 and 536 nm. The TEM analysis confirmed the spherical, triangular, and hexagonal shapes of AuNPs. In aqueous solutions, for sensing Sr^{2+}, Ca^{2+}, Zn^{2+}, and Cu^{2+} ions, these AuNPs could be utilized as plasmon resonance and colorimetric sensors despite their different shapes and sizes. In several metal ions, the developed biogenic AuNPs were also observed for the sensitivity studies. The solid biogenic cellulosic biocomposites/AuNPs were used as a fast, portable, and reliable colorimetric sensor. Though virtuous adsorbent materials of metal ions corresponded to these biocomposites [41].

Similarly, the functionalized AuNPs were synthesized using a sonochemical process and pulsed laser irradiation in the presence of lignin matrixes. Various analytical techniques were used for the characterization of the resulting lignin functionalized AuNPs (L-Auf NPs), i.e., X-ray diffraction, high-resolution TEM, FTIR, and UV–vis spectroscopy. The L-Auf NPs showed extreme selectivity for the colorimetric detection of Pb^{2+} ions in mixed metal ion solution within a short time interval. The detection process has been confirmed by the prominent color change from red wine to purple of L-Auf NPs after addition of Pb^{2+}

ions. The developed NPs showed 1.8 μM of LOD in the linear range of 0.1–1 mM. Therefore, the developed sensor is considered to be vital for the detection of Pb^{2+} ions in aqueous samples [42].

Based on SPR, optical sensors were used to detect toxic heavy metals in solutions. Nanostructured thin films were produced with active layers to progress the sensitivity of SPR sensors. To measure and compare the response to Pb^{2+} for SPR sensors, the thin films of Au/CS films and gold–chitosan–graphene oxide nanostructures (Au/CS/GO) were advanced. The field emission scanning electron microscopy (FESEM) analysis showed that the developed Au/CS/GO's is a film composed of nanosheets with rough and wrinkled surfaces. The effective incorporation of GO in the prepared films showed by the XRD analysis.

The results showed that the Au/CS films were suggestively smooth than the Au/CS/GO films, which confirmed the AFM analysis, demonstrating that the Au/CS/GO films had a root mean square (RMS) roughness of 28.38 nm. Upon adding a 5 ppm solution of Pb^{2+} ion, the Au/CS films and Au/CS/GO films showed progressive SPR sensitivity minimum to 0.77600 and 1.11200 ppm. This enhancement of the SPR response was observed in these films because of the strong covalent bonding between CS and GO. The Au/CS/GO films show potential applications for detecting heavy metals [43].

The usage of the extract of grapefruit (*Citrus paradisi*) as the capping and reducing agent for the synthesis of AuNPs has been done via a fast, biosynthetic, entirely biological approach in a single step. The aqueous extract of grapefruit quickly reduces the Au^{3+} ions, corresponding to enormously crystalline colloidal and stable AuNPs. In the UV–vis spectra, the SPR peak at 544 nm confirms the formation of AuNPs. In aqueous media, for detecting Ca^{2+}, Pb^{2+}, Zn^{2+}, Cu^{2+}, and Hg^{2+} ions, the biogenic AuNPs showed a virtuoso performance and selectivity as plasmonic, fluorescent, and naked-eye sensors. Three methods have been utilized out of them, and the fluorescent sensors displayed virtuous outcomes precisely with Ca^{2+}, Cu^{2+}, and Pb^{2+}. For the detection of Cu^{2+} ions, plasmonic and fluorescent approaches were utilized [44].

At 45 °C, by mixing chitosan with carbon disulfide and ammonia, chitosan dithiocarbamate (CSDTC) derivative was successfully synthesized. After that, AuNPs solution mixed with CSDTC (40L, 1.0 mg mL^{-1}) to get the AuNPs functionalized with chitosan dithiocarbamate (CSDTC-AuNPs). After adding Cd^{2+} ions to developed CSDTC-AuNPs, the particles got aggregated quickly at PBS pH 8.0. Therefore, giving a change in color from red to blue, a red-shift from 523 to 685 nm observed in the SPR peak of CSDTC-Au NPs. The detection probe showed 63 nM of LOD toward Cd^{2+} ions. CSDTC-AuNPs were applied as a potential sensor in real water samples to determine Cd^{2+} ion, for example, tap, industrial effluent, river, canal water, and drinking water [45].

In ultrasound radiation, to detect heavy metal ions, palm oil fronds extracts (POFE) have been utilized to synthesize AuNPs. The POFE is accessible in abundance as waste material in Asia and African countries. The reduction of Au^{3+} to Au^0 was possibly made thru the functional groups present in the POFE, which functions as a capping and reducing agent. Various analytical tools have been used to characterize prepared AuNPs, for example, FTIR, FESEM, UV–vis, XRD, and DLS. Their FTIR spectra confirmed the coating of phenolic and alkynes complexes on the AuNPs, which displays a viable role of biomolecules for the effective stabilization of AuNPs. FESEM shows the surface morphology of the developed AuNPs. The XRD patterns at (111), (200), (220), (311), and (222) showed the fcc crystal structures of as-synthesized material. The biosynthesized AuNPs selectively detect Cr metal when its concentration is augmented from 1 to 50 ppm, further confirmed using UV–vis at room temperature with a rapid decrease of intensity. Simultaneously, almost all other metal ions are detected at raised temperatures (45 °C–50 °C) [46].

To synthesize Au and Ag NPs biologically, Ficus retusa was utilized as stabilizing and reducing agent with controllable shape and size and high dispersion stability. The distribution of the size of the MNPs and tuning of the particle size can be done thru the optimizing of reaction parameters extract quantity, contact time, pH value, and metal concentration. The SPR of the MNPs was observed using UV–vis to follow the spectral profile changes due to various actions. Depending on the diverse reaction parameters, the SPR alters between 522 and 554 nm for AuNPs, whereas, for silver, it alters among 400 and 432 nm, correspondingly. The fruitful development of spherical gold (10–25 nm) and silver (15 nm) NPs was confirmed through TEM and AFM with narrow size distribution. The key role phenolic compounds are as stabilizing and reducing agents for metal ions resulted from FTIR. The colorimetric sensitivity of Au and AgNPs was also studied [47].

3.4 CONCLUSIONS, FUTURE PERSPECTIVES, AND CHALLENGES

Nowadays, research toward the development of biological sensors fascinating significant action and attention. This comprises AuNPs and AgNPs in recognition of toxic metal ions, and the current improvement in the synthesis of biological AuNP, as colorimetric sensors for environmental sensing. Earlier, in the developing field of AgNPs and AuNP-plasmonics, fabrication and synthesis of biological AgNPs and AuNP have directed to an explosion of reports and efforts equally. In producing novel nano-sized sensors for biological imaging and sensing, it has made substantial improvement. Due to the compatible nature of AgNPs and AuNPs with portable devices, which may be organized in the area for onsite sample screening, AgNPs and AuNP-based sensors offer auspicious methods for advantages over traditional instrumental analyses and sensitive and selective studies. This portable device can be manufactured to measure untreated field-collected water samples with increased sensitivity and rapid, quantitative, and multiplexed examination with the development of microfluidic sample handling units. The developed green AgNPs and AuNPs sensors offer effective screening methods based on ultrasensitive and high-throughput detection technology for various ecological analytes. Soon, this method will surely discover extensive benefits in water safety and quality monitoring. Two lines of research are actively exploring researchers globally:

- New sensing platforms and instruments are being exploited and intended based on their growing knowledge bases to offer extraordinary resolution and sensitivity for biological and chemical imaging and sensing; and
- At the nanoscale new materials, to increase our consideration of plasmonic phenomena, new models and theories are being investigated and fabricated, for example, anisotropic-shaped AgNPs, Au alloy NPs, or AuNPs.

Hence, to yield next generation nanosensors, these two lines of research are merging. Though environmental monitoring should be considered before being realistic tools by these green developed AgNPs and AuNPs, numerous serious problems still need to be addressed. The long-term health effects of nanomaterials, which are vital for environmental applications, still need to be characterized, and a high quantity analysis needs to be developed. However, soon, we expect that a novel, vital insight into the AgNPs and AuNP-based nanotechnology would be developed, resulting in more powerful nanosensors for various applications by increased collaboration between scientists from different disciplines.

REFERENCES

1. Arora, Monu, Bala Kiran, Shweta Rani, Anchal Rani, Barinder Kaur, and Neeraj Mittal. "Heavy metal accumulation in vegetables irrigated with water from different sources." *Food Chemistry* 111, no. 4 (2008): 811–815. doi:10.1016/j.foodchem.2008.04.049.
2. Khlifi, Rim, and Amel Hamza-Chaffai. "Head and neck cancer due to heavy metal exposure via tobacco smoking and professional exposure: A review." *Toxicology and Applied Pharmacology* 248, no. 2 (2010): 71–88. doi:10.1016/j.taap.2010.08.003.
3. Jaishankar, Monisha, Tenzin Tseten, Naresh Anbalagan, Blessy B. Mathew, and Krishnamurthy N. Beeregowda. "Toxicity, mechanism and health effects of some heavy metals." *Interdisciplinary Toxicology* 7, no. 2 (2014): 60. doi:10.2478/intox-2014-0009.
4. Poornima, Velswamy, Vincent Alexandar, S. Iswariya, Paramasivan T. Perumal, and Tiruchirappalli Sivagnanam Uma. "Gold nanoparticle-based nanosystems for the colorimetric detection of Hg^{2+} ion contamination in the environment." *RSC Advances* 6, no. 52 (2016): 46711–46722. doi:10.1039/C6RA04433F.
5. Lin, Wen-Chi, Zhongrui Li, and Mark A. Burns. "A drinking water sensor for lead and other heavy metals." *Analytical Chemistry* 89, no. 17 (2017): 8748–8756. doi:10.1021/acs.analchem.7b00843.
6. Gulu, Ilker. "Determination of heavy metal accumulation in plant samples by spectrometric techniques in Turkey." *Applied Spectroscopy Reviews* 50, no. 2 (2015): 113–151. doi:10.1080/05704928.2014.935981.
7. Hoang, Chung V., Makiko Oyama, Osamu Saito, Masakazu Aono, and Tadaaki Nagao. "Monitoring the presence of ionic mercury in environmental water by plasmon-enhanced infrared spectroscopy." *Scientific Reports* 3, no. 1 (2013): 1–6. doi:10.1038/srep01175.
8. Li, Ming, Honglei Gou, Israa Al-Ogaidi, and Nianqiang Wu. "Nanostructured sensors for detection of heavy metals: A review." *ACS Sustainable Chemistry & Engineering* 1, no. 7 (2013): 713–723. doi:10.1021/sc400019a.

9. Yuan, Xun, Magdiel Inggrid Setyawati, Audrey Shu Tan, Choon Nam Ong, David Tai Leong, and Jianping Xie. "Highly luminescent silver nanoclusters with tunable emissions: Cyclic reduction–decomposition synthesis and antimicrobial properties." *NPG Asia Materials* 5, no. 2 (2013): e39–e39. doi:10.1038/am.2013.3.

10. Kaur, Balwinder, Rajendra Srivastava, and Biswarup Satpati. "Ultratrace detection of toxic heavy metal ions found in water bodies using hydroxyapatite supported nanocrystalline ZSM-5 modified electrodes." *New Journal of Chemistry* 39, no. 7 (2015): 5137–5149. doi:10.1039/C4NJ02369B.

11. Hung, Yu-Lun, Tung-Ming Hsiung, Yi-You Chen, Yu-Fen Huang, and Chih-Ching Huang. "Colorimetric detection of heavy metal ions using label-free gold nanoparticles and alkanethiols." *The Journal of Physical Chemistry C* 114, no. 39 (2010): 16329–16334. doi:10.1021/jp1061573.

12. Aragay, Gemma, Josefina Pons, and Arben Merkoçi. "Enhanced electrochemical detection of heavy metals at heated graphite nanoparticle-based screen-printed electrodes." *Journal of Materials Chemistry* 21, no. 12 (2011): 4326–4331. doi:10.1039/C0JM03751F.

13. De Acha, Nerea, César Elosúa, Jesús M. Corres, and Francisco J. Arregui. "Fluorescent sensors for the detection of heavy metal ions in aqueous media." *Sensors* 19, no. 3 (2019): 599. doi:10.3390/s19030599.

14. Borah, Sandhya B. D., Tanujjal Bora, Sunandan Baruah, and Joydeep Dutta. "Heavy metal ion sensing in water using surface plasmon resonance of metallic nanostructures." *Groundwater for Sustainable Development* 1, no. 1–2 (2015): 1–11. doi:10.1016/j.gsd.2015.12.004.

15. Narayanan, Kannan Badri, and Natarajan Sakthivel. "Synthesis and characterization of nano-gold composite using *Cylindrocladium floridanum* and its heterogeneous catalysis in the degradation of 4-nitrophenol." *Journal of Hazardous Materials* 189, no. 1–2 (2011): 519–525. doi:10.1016/j.jhazmat.2011.02.069.

16. Yurkov, Andrey M., Martin Kemler, and Dominik Begerow. "Species accumulation curves and incidence-based species richness estimators to appraise the diversity of cultivable yeasts from beech forest soils." *PLoS One* 6, no. 8 (2011): e23671. doi:10.1371/journal.pone.0023671.

17. Marchiol, Luca. "Synthesis of metal nanoparticles in living plants." *Italian Journal of Agronomy* (2012): e37–e37. doi:10.4081/ija.2012.e37.

18. Iravani, Siavash. "Bacteria in nanoparticle synthesis: Current status and future prospects." *International Scholarly Research Notices* 2014 (2014). doi:10.1155/2014/359316.

19. Singh, Jagpreet, Tanushree Dutta, Ki-Hyun Kim, Mohit Rawat, Pallabi Samddar, and Pawan Kumar. "'Green' synthesis of metals and their oxide nanoparticles: Applications for environmental remediation." *Journal of Nanobiotechnology* 16, no. 1 (2018): 1–24. doi:10.1186/s12951-018-0408-4.

20. Ravi, Selvan Sukanya, Lawrence Rene Christena, Nagarajan SaiSubramanian, and Savarimuthu Philip Anthony. "Green synthesized silver nanoparticles for selective colorimetric sensing of Hg^{2+} in aqueous solution at wide pH range." *Analyst* 138, no. 15 (2013): 4370–4377. doi:10.1039/C3AN00320E.

21. Annadhasan, M., T. Muthukumarasamyvel, V. R. Sankar Babu, and N. Rajendiran. "Green synthesized silver and gold nanoparticles for colorimetric detection of Hg^{2+}, Pb^{2+}, and Mn^{2+} in aqueous medium." *ACS Sustainable Chemistry & Engineering* 2, no. 4 (2014): 887–896. doi:10.1021/sc400500z.

22. Ahmed, Muhammad Ahad, Najmul Hasan, and Shaikh Mohiuddin. "Silver nanoparticles: Green synthesis, characterization, and their usage in determination of mercury contamination in seafoods." *International Scholarly Research Notices* 2014 (2014). doi:10.1155/2014/148184.

23. Devnani, H., and S. P. Satsangee. "Green gold nanoparticle modified anthocyanin-based carbon paste electrode for voltammetric determination of heavy metals." *International Journal of Environmental Science and Technology* 12, no. 4 (2015): 1269–1282. doi:10.1007/s13762-014-0497-z.

24. Luo, Yuqiong, Suqin Shen, Jiwen Luo, Xiaoying Wang, and Runcang Sun. "Green synthesis of silver nanoparticles in xylan solution via Tollens reaction and their detection for Hg^{2+}." *Nanoscale* 7, no. 2 (2015): 690–700. doi:10.1039/C4NR05999A.

25. Annadhasan, M., and N. Rajendiran. "Highly selective and sensitive colorimetric detection of Hg^{2+} ions using green synthesized silver nanoparticles." *RSC Advances* 5, no. 115 (2015): 94513–94518. doi:10.1039/C5RA18106B.

26. Alam, Md Niharul, Anirban Chatterjee, Sreeparna Das, Shaikh Batuta, Debabrata Mandal, and Naznin Ara Begum. "Burmese grapefruit juice can trigger the "logic gate"-like colorimetric sensing behavior of Ag nanoparticles towards toxic metal ions." *RSC Advances* 5, no. 30 (2015): 23419–23430. doi:10.1039/C4RA16984K.

27. Annadhasan, M., J. Kasthuri, and N. Rajendiran. "Green synthesis of gold nanoparticles under sunlight irradiation and their colorimetric detection of Ni^{2+} and Co^{2+} ions." *RSC Advances* 5, no. 15 (2015): 11458–11468. doi:10.1039/C4RA14034F.

28. May, B. M., and Oluwatobi S. Oluwafemi. "Sugar-reduced gelatin-capped silver nanoparticles with high selectivity for colorimetric sensing of Hg^{2+} and Fe^{2+} ions in the midst of other metal ions in aqueous solutions." *International Journal of Electrochemical Science* 11 (2016): 8096–8108. doi:10.20964/2016.09.29.

29. Tirado-Guizar, Antonio, Geonel Rodriguez-Gattorno, Francisco Paraguay-Delgado, Gerko Oskam, and Georgina E. Pina-Luis. "Eco-friendly synthesis of egg-white capped silver nanoparticles for rapid, selective, and sensitive detection of Hg^{2+}." *MRS Communications* 7, no. 3 (2017): 695. doi:10.1557/mrc.2017.74.

30. Firdaus, M., S. Andriana, W. Alwi, E. Swistoro, A. Ryan, and A. Sundaryono. "Green synthesis of silver nanoparticles using Carica Papaya fruit extract under sunlight irradiation and their colorimetric detection of mercury ions." *In Journal of Physics: Conference Series* 817, no. 1 (2017): 012029. IOP Publishing. doi:10.1088/1742-6596/817/1/012029.

31. Zhang, Bin, Jiang Chen, Han Zhu, TingTing Yang, Meiling Zou, Ming Zhang, and MingLiang Du. "Facile and green fabrication of size-controlled AuNPs/CNFs hybrids for the highly sensitive simultaneous detection of heavy metal ions." *Electrochimica Acta* 196 (2016): 422–430. doi:10.1016/j.electacta.2016.02.163.

32. Cheon, Ja Young, and Won Ho Park. "Green synthesis of silver nanoparticles stabilized with mussel-inspired protein and colorimetric sensing of Pb^{2+} and Cu^{2+} ions." *International Journal of Molecular Sciences* 17, no. 12 (2016): 2006. doi:10.3390/ijms17122006.

33. Karthiga, D., and Savarimuthu Philip Anthony. "Selective colorimetric sensing of toxic metal cations by green synthesized silver nanoparticles over a wide pH range." *RSC Advances* 3, no. 37 (2013): 16765–16774. doi:10.1039/C3RA42308E.

34. Ihsan, Muhammad, Abdul Niaz, Abdur Rahim, Muhammad Iqbal Zaman, Muhammad Balal Arain, Tehmina Sharif, and Memoona Najeeb. "Biologically synthesized silver nanoparticle-based colorimetric sensor for the selective detection of Zn^{2+}." *RSC Advances* 5, no. 111 (2015): 91158–91165. doi:10.1039/C5RA17055A.

35. Alzahrani, Eman. "Colorimetric detection based on localized surface plasmon resonance optical characteristics for sensing of mercury using green-synthesized silver nanoparticles." *Journal of Analytical Methods in Chemistry* 2020 (2020). doi:10.1155/2020/6026312.

36. Sebastian, Maria, Archana Aravind, and Beena Mathew. "Green silver-nanoparticle-based dual sensor for toxic Hg^{2+} ions." *Nanotechnology* 29, no. 35 (2018): 355502. doi:10.1088/1361-6528/aacb9a.

37. Maiti, Swarnali, Gadadhar Barman, and Jayasree Konar Laha. "Detection of heavy metals (Cu^{2+}, Hg^{2+}) by biosynthesized silver nanoparticles." *Applied Nanoscience* 6, no. 4 (2016): 529–538. doi:10.1007/s13204-015-0452-4.

38. Kumar, Vijay, Devendra K. Singh, Sweta Mohan, Daraksha Bano, Ravi Kumar Gundampati, and Syed Hadi Hasan. "Green synthesis of silver nanoparticle for the selective and sensitive colorimetric detection of Hg^{2+} ion." *Journal of Photochemistry and Photobiology B: Biology* 168 (2017): 67–77. doi:10.1016/j.jphotobiol.2017.01.022.

39. Ankamwar, Balaprasad. "Green synthesized unmodified silver nanoparticles as colorimetric sensors for the selective detection of Hg^{2+} ions." *Chemical Science Review and Letters* 5, no. 19 (2016): 312–317. ISSN 2278-6783.

40. Ahmed, Farid, Humaira Kabir, and Hai Xiong. "Dual colorimetric sensor for Hg^{2+}/Pb^{2+} and an efficient catalyst based on silver nanoparticles mediating by the root extract of *Bistorta amplexicaulis*." *Frontiers in Chemistry* 8 (2020). doi:10.3389/fchem.2020.591958.

41. Hoyos, Silva-De, E. Luisa, Victor Sánchez-Mendieta, Alfredo R. Vilchis-Nestor, Miguel A. Camacho-López, Jésica Trujillo-Reyes, and Miguel Avalos-Borja. "Plasmonic sensing of aqueous-divalent metal ions by biogenic gold nanoparticles." *Journal of Nanomaterials* 2019 (2019). doi:10.1155/2019/9846729.

42. Yu, Yiseul, Shreyanka Shankar Naik, O. Yewon, Jayaraman Theerthagiri, Seung Jun Lee, and Myong Yong Choi. "Lignin-mediated green synthesis of functionalized gold nanoparticles via pulsed laser technique for selective colorimetric detection of lead ions in aqueous media." *Journal of Hazardous Materials* (2021): 126585. doi:10.1016/j.jhazmat.2021.126585.

43. Lokman, Nurul Fariha, Ahmad Ashrif A. Bakar, Fatihah Suja, Huda Abdullah, Wan Baihaqi Wan Ab Rahman, Nay-Ming Huang, and Mohd Hanif Yaacob. "Highly sensitive SPR response of Au/chitosan/graphene oxide nanostructured thin films toward Pb^{2+} ions." *Sensors and Actuators B: Chemical* 195 (2014): 459–466. doi:10.1016/j.snb.2014.01.074.

44. Silva-De Hoyos, Luisa E., Victor Sanchez-Mendieta, Miguel A. Camacho-Lopez, Jesica Trujillo-Reyes, and Alfredo R. Vilchis-Nestor. "Plasmonic and fluorescent sensors of metal ions in water based on biogenic gold nanoparticles." *Arabian Journal of Chemistry* 13, no. 1 (2020): 1975–1985. doi:10.1016/j.arabjc.2018.02.016.

45. Mehta, Vaibhavkumar N., Hirakendu Basu, Rakesh Kumar Singhal, and Suresh Kumar Kailasa. "Simple and sensitive colorimetric sensing of Cd^{2+} ion using chitosan dithiocarbamate functionalized gold nanoparticles as a probe." *Sensors and Actuators B: Chemical* 220 (2015): 850–858. doi:10.1016/j.snb.2015.05.105.

46. Usman, Adamu Ibrahim, and Azlan Abdul Aziz. "Photometric detection of heavy metals using biosynthesized gold nanoparticles." In *Solid State Phenomena*, vol. 301, pp. 118–123. Trans Tech Publications Ltd, 2020. doi:10.4028/www.scientific.net/SSP.301.118.

47. Zayed, Mervat F., Wael H. Eisa, Salah M. El-Kousy, Walaa K. Mleha, and Nermeen Kamal. "Ficus retusa-stabilized gold and silver nanoparticles: Controlled synthesis, spectroscopic characterization, and sensing properties." *Spectrochimica Acta Part A: Molecular and Biomolecular Spectroscopy* 214 (2019): 496–512. doi:10.1016/j.saa.2019.02.042.

4 Chemically Functionalized Silver and Gold Nanostructures as SPR Based Sensors for the Detection of Toxic Metal Ions in Aqueous Media

4.1 INTRODUCTION

Significant considerations have been received in the last decade as colorimetric sensors to detect heavy metal ions. Because of the colorimetric properties of metallic NPs of noble metals. Among the MNPs, the most stable MNPs are AuNPs. Gold is essential to be applied in devices and nanotechnologies due to its resistance to surface oxidation and chemical inertness [1,2]. Although AgNPs, which are less stable than AuNPs, display the maximum efficacy of plasmon excitation. In contrast to AuNPs of the same size, AgNPs have a 100-fold greater molar extinction coefficient, and they have improved visibility and increased sensitivity (because of the alterations in optical brightness) [3].

In contrast to a particle of any identified inorganic or organic chromophore of the same dimension, a single AgNP interacts more efficiently with light. Simultaneously, the usage of AgNPs as colorimetric sensors fascinated much attention, owing to their cost-effective nature, in contrast to AuNPs [4,5]. The surface modification shows a vital role in augmenting the analytical applicability and stability of AgNPs, which is confirmed by the vulnerability to oxidation of the silver surface. Colloidal gold was utilized for a long time for attractive requests to synthesize colored ceramics and ruby glass. Contrary to the restoring force of positive nuclei, the EM radiation (incident photon frequency) resonates with the surface electron's natural frequency because of a plasmon-absorption

DOI: 10.1201/9781003128281-4

band's attendance and the unique optical properties and bright colors noble-metal particles, such as silver and gold. This effect is indicated through LSPR [6].

4.2 AuNPs AND AgNPs FOR THE COLORIMETRIC DETECTION OF METAL IONS

The AgNPs and AuNPs-based colorimetric methods overcome some limits of the utmost conventional approaches and do not acquaint with length and complicated procedures. AgNPs and AuNPs correspond to colorimetric behavior due to their very sensitive nature toward the environment's dielectric nature (refractive index). These colors' detection processes were done thru the naked eye based on interparticle interactions (aggregation), NP surface modifications, self-assembly, and a refractive index of the medium. The redshift in the absorption spectra of AuNPs occurs due to the refractive index of the medium [7]. Due to these visual changes, imaging and sensing of a wide range of analyte functionalized AgNPs and AuNPs are widely explored as probes, such as metal ions [8,9]. Optical properties, aggregation of NP, and their size are controlled in solution through the SPR of AgNPs and AuNPs. The color of small AuNPs is red, which turned to blue upon aggregation, whereas the color of AgNPs is yellow, which could change to red upon aggregation. This alteration in color results from the surface plasmon resonance coupling among particles. In the UV–vis spectrum, the effect of plasmonic coupling upsurges as the distance among particles declines, ensuing in color changes besides the displacement of the LSPR. The nature of functional groups and the ligands functionalized into the NP surface can control the NPs' selectivity, affinity, and aggregation toward metal ions. The precise molecular detection of ligands joining the NPs thru metal ions works as a bridge. Similarly, according to the soft and hard acids and bases theory, the selectivity of metal ions is defined, establishes that ligands rich in sulfur and nitrogen for heavy metal and transition ions and toward alkaline earth metal ions the oxygen atoms rich ligands have a robust competence [10]. Table 4.1 comprises different chemically synthesized AuNPs for the detection of metals colorimetrically and electrochemically.

4.3 CHEMICALLY SYNTHESIZED AuNPs TO DETECT HEAVY METALS

In a similar study, for detecting various heavy metal ions, Mn, Cd, Fe, Cr, Pb, Co, Hg, Ti, and Zn, the ascorbate-, glutathione (GSH), and citrate-functionalized AuNPs have been developed. The AuNPs-based removal/sensing of heavy metals results from these sorts of interactions among the surface moieties of functionalizing agents and heavy metal ions. The spectroscopic signals and color of functionalized AuNPs have been evaluated thru analyzing the variations in these interactions. The attained outcomes were also validated and compared by the computational studies. With E force values of -14.0 and $-23.4\,kJ\,mol^{-1}$ for Mn and Cr, GSH-AuNPs and citrate-AuNPs exhibited high selectivity, correspondingly.

TABLE 4.1
Detection of Metals Using AgNPs by the Colorimetric Method

Nanomaterials	Size	Detection Method	Analytes	LOD	References
(AuNPs-GN-Cys) composites		Electrochemical	Pb^{2+} Cd^{2+}	0.05 µg L^{-1} 0.10 µg L^{-1}	[11]
GQDs functionalized AuNPs	50 nm	Electrochemical	Cu^{2+}, Hg^{2+}	0.05 nM 0.02 nM	[12]
3D graphene-framework/Bi NPs	–	Electrochemical	Zn^{2+} Pb^{2+} Cd^{2+}	300 µg L^{-1} 0.02 µg L^{-1} 0.05 µg L^{-1}	[13]
Tween-20 and gelatin AgNPs	17.54 nm 9.68 nm	Colorimetric	Hg^{2+}	0.13 mg L^{-1} 0.45 mg L^{-1}	[14]
PVA-AgNPs	–	Colorimetric	Hg^{2+}	10 ppb	[15]

Similarly, with the E force value of −19.6 kJ mol^{-1}, the ascorbate-AuNPs exhibited sensitivity for multiple ions, such as Mn, Fe, and Cd. To display precise interactions among heavy metal ions and functionalized AuNPs, a detailed analysis concentrates on the interaction energy values, ionic sizes, and electrostatic charges. The respective mechanisms of interaction among functionalized AuNPs and heavy metal ions have been discovered [16].

Through drinking water, surplus exposure to toxic metal ions causes life-threatening diseases and poses high risks to human health. Therefore, it is vital to synthesizing a low-cost, sensitive, and rapid method for detecting metal ions in water. For colorimetric detection of Al^{3+} and Pb^{2+} ions, lab-on-a-chip instrumentation based on gold NP probes developed in the water system. The sensing of metal ions colorimetrically with high selectivity and low LOD is conducted via single-step assays. A handheld colorimetric reader and a custom-made microwell plate have been designed to quantify the signal and implement the assay's readout. The LOD was found 89 ppb for Al^{3+}, and 30 ppb for Pb^{2+} ions confirmed by the calibration experiments on a developed portable system comparable to benchtop analytical spectrometers. For resource-poor settings and in-field water quality monitoring more conveniently and economically, the detection probe promises an effective platform for analyzing metal ions [17]. Figure 4.1 demonstrates the mechanism for the detection of metal ions colorimetrically using chemically functionalized AuNPs.

For the Hg^{2+} ion detection, a new label-free, selective and sensitive optical technique based on the SPR has been established through GNRs functionalized with cyano (−CN) ligand. Before detecting Hg^{2+} ions on the GNRs surface, the functionalized GNRs were used with poly(2-aminobenzonitrile) (P2ABN), which is a −CN group-containing polymer. Two methods were used for the functionalization of the GNRs. The first method, a porous silica network, was containing P2ABN used to modify the surface of GNR for the synthesis of GNR@silica-CN. The second method was used to develop GNR@P2ABN thru the functionalization

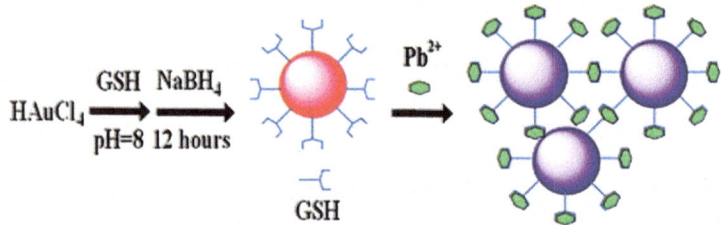

FIGURE 4.1   A mechanism for colorimetric detection of lead ions using GSH-GNPs. (Reprinted with permission from Ref. [20]. Copyright 2010 American Chemical Society.)

of GNR with P2ABN. By monitoring the SPR of GNRs, reducing Hg^0 atoms by AA, and the interaction of the –CN groups in P2ABN with Hg^{2+} ions, the detection method of Hg^{2+} ion includes the detection of Hg^{2+} ions on GNR@P2ABN surface or GNR@silica-CN.

In contrast to GNR@P2ABN, to reach the GNR surface, the porous network in GNR@silica-CN permits effective movement of Hg^0 and results in effective amalgamation. Resultant, 50 nM to 5 mM of eclectic concentration range, the GNR@silica-CN displays a substantial alteration for Hg^{2+} ions in the SPR of GNRs. For ultra-low levels of Hg ion detection and monitoring, the GNR@silica-CN showed excellent potential for Hg^{2+} ions having the lowest detection limit up to 1 ppb. The GNR@silica-CN displays the high specificity for detecting Hg^{2+} ions in water samples, confirmed by the sensitivity check-in mix metal ion solution. The developed detection probe also showed outstanding efficiency in detecting Hg^{2+} ions in spiked water samples (pond). In the future, the suitability of sensitive, fast, and simple methods was predicted for environmental monitoring [18]. Figure 4.2 shows the selectivity of the detection probe in a mixture of diverse metal ion solutions.

An optical fiber sensor has been developed in another study to detect Pb^{2+} ions in a water sample. The developed U-shaped optical fiber was selected for the detection process because of a point of care capability and higher sensitivity. In the sensing probe, AuNPs functionalized with oxalic acid were coated. Intensity, power, and output voltage were measured using an optical photodetector consistent with a disparity in lead ion concentration. The linear fit curves results showed R^2 values of 0.98, 0.98, and 0.99 and LOD 2.10, 1.75, and 1.87 ppb in the linear range of 1–20 ppb respective to intensity response, power, and voltage curve. The developed probed selectivity detects lead ions and has a negligible effect on this sensing element in mixed metal ion solution. Instead of the spectrometer, the developed sensor probe was cost-effective and compact due to adopting a U-shaped probe with an optical detector [19].

GSH functionalized AuNPs (GSH-GNPs) have been developed as sensitive, low-cost, and facile colorimetric sensing techniques for detecting Pb^{2+} ions. The synthesized GSH-GNPs have been checked for selectivity and sensitivity studies. After adding 1 M NaCl aqueous solution, in the presence of a Pb^{2+} ion, the GSH-GNPs can be persuaded for the aggregation instantly. The LOD was found

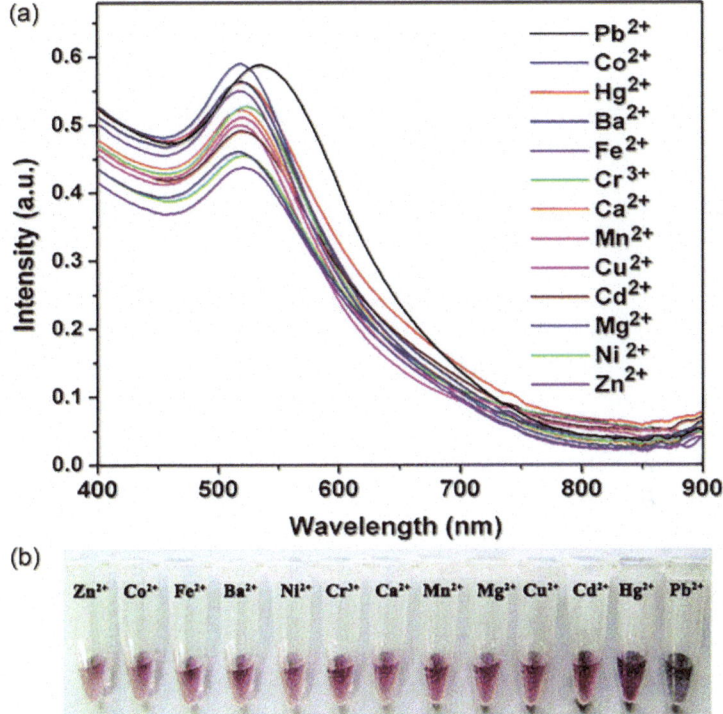

FIGURE 4.2 (a) Selectivity of a probe by measuring SPR using UV–vis absorption spectra and (b) Optical images of GSH-GNPs incubation with metal ions for 20 min containing 50 μM other metal ions compared with 20 μM Pb²⁺. (Reprinted with permission from Ref. [20]. Copyright 2010 American Chemical Society.)

100 nM for Pb^{2+} ions, which was confirmed by the colorimetric response of GNPs using naked eyes or a UV–vis spectrophotometer. In contrast to other metal ions such as Hg^{2+}, Mg^{2+}, Ni^{2+}, Cr^{3+}, Co^{2+}, Zn^{2+}, Cu^{2+}, Mn^{2+}, Cd^{2+}, Ca^{2+}, Fe^{2+}, and Ba^{2+}, the GSH-GNPs exhibited outstanding selectivity toward Pb^{2+} ions, which correspond to protuberant color alteration. For on-site and real-time detection of Pb^{2+} ions, this method provided a practical and straightforward colorimetric sensor. Primarily, this probe with high sensitivity and low interference was also utilized to detect Pb^{2+} ions in the lake water samples [20].

In a similar study, divalent mercuric ion (Hg^{2+}) is of utmost hazardous and extensively diffused biological contaminants as an exceedingly toxic heavy metal ion. The AuNPs functionalized by 4-mercaptopyridine (4-MPY) have been considered for detecting Hg^{2+} ions as a signal amplification tag. The developed sensor is based on the SPR effect, which takes advantage of an inexpensive, portable, and simple fiber-optic sensor. The sensor was developed by introducing AuNPs/4-MPY through self-assembling 4-MPY on Au film surfaces and capturing Hg^{2+} ions. The detection technique is based on the coordination among nitrogen in the pyridine moiety and Hg^{2+} ions. In an aqueous solution, highly sensitive Hg^{2+} sensing is augmented thru changes in SPR wavelength by the coupling among localized SPR and metal ions. In contrast to the other metal ions, the sensor selectivity detects Hg^{2+} ion, which showed LOD 8 nM toward Hg^{2+} ion under optimal conditions. Moreover, the developed probe also experimented with detecting mercury in real water samples [21].

The SPR-based AuNPs functionalized by diglycolic acid have been reported to sense Cr^{6+} ions at low concentrations of Cr^{6+} ions in aqueous solutions. The capped diglycolic acid persuaded the NP probes' aggregation upon encountering them because of its excellent affinity for Cr^{6+} ions, which was further confirmed by the enlarged size of the NPs and the noticeable red shifting of the SPR peak. The probe displayed the 0.32 ppb LOD under optimized conditions, with 0.32 ppb to 0.1 ppm of linear detection scale for Cr^{6+} ions. Among SPR-based methods, the developed sensor probe showed the lowest LOD reported for Cr^{6+} detection [22].

The surface functionalization possibility with many ligands is the key benefit of AuNPs. Introducing ligands through a coordination reaction on the surface of AuNPs offers the opportunity to interact with metal ions and stabilize the NPs in solution. For metal coordination and anchoring it onto the gold surface, the molecules are designed with bifunctional groups. Therefore, for the selective and sensitive sensing of a precise metal ion, the coupling among the supramolecular chemistry and NPs is perfect for attaining the required aim. A varying level of K^+ concentration shows many ailments, as with additional metals. Chen et al. [23] established a plasmonic assay to sense potassium selectively using thiolated aptamers functionalized AuNPs. The detection of K^+ ions occurs via the Au-S bond where thiolated aptamers were linked to the AuNP surface for the detection process. Persuading NPs' aggregation, the K^+ binds to the G-rich nucleic acid, forming a G4 structure [23].

Similarly, lactose functionalized AuNPs have been utilized for the first time to sense Ca^{2+} ions. The aggregation of NPs leads to colorimetric changes after the addition of Ca^{2+}, which induces communication among the carbohydrate moieties. Two colorimetric approaches were reported for Ca^{2+} ion detection based on cytidine triphosphate (CTP) [24] and calsequestrin (CSQ) [25] functionalized AuNPs. The Ca^{2+} ions handled a colorimetric change. Furthermore, CSQ functionalized AuNPs system formed aggregates of 13 nm. Here, a red to blue color change was obtained when the Ca^{2+} ion concentration upsurges from 10 μM to 0.01–1 mM and the polymerization of CSQ-AuNPs was detected. An alteration of color and a redshift occurred from 520 to 600 nm to detect Ca^{2+} ions thru the CTP@AuNPs system via aggregation. The first technique has a lower detection limit of 0.01 mM among both the methods, which have a similar detection technique via NP aggregation.

Zhang et al. [26] synthesized a colorimetric probe of AuNPs functionalized with 2-mercaptosuccinic acid (MSA) to detect alkaline earth metal ions. The developed probe is able to detect Ba^{2+}, Sr^{2+}, and Ca^{2+} metal ions concurrently. The thiol group of MSA molecules handled the binding to the surface gold surface. The size of synthesized citrate-capped AuNPs in this system was 13 nm.

Corresponding to NP aggregation via a bathochromic shift of the plasmon band, forming a strong chelator among MSA molecules occurs after adding Ba^{2+}, Sr^{2+}, and Ca^{2+}. The results showed that the color change was observed thru the linkage of the two adjacent MSA molecules with the four carboxylic oxygen atoms [27].

4.4 CHEMICALLY SYNTHESIZED AgNPs TO DETECT HEAVY METALS

A simple photochemical process was used to synthesize silver NPs wrapped with reduced GO (Ag@rGO), where the GO nanosheets were utilized as both stabilizing and reducing agents. The synthesized Ag@rGO nanohybrids were being used as LSPR based sensors to sense carbaryl and Cr^{6+}. The outcome for the Cr^{6+} exhibited a change in color from yellow to colorless. They found the linear dependence of the concentration to absorption intensity in 0.1–25 µM of concentration range for Cr^{6+}. Consistent with the yellow to purple color change in the calibration curve, a high linearity level was attained for the carbaryl between 0.1 and 50 µM. The LODs were 42 and 31 nM for carbaryl and Cr^{6+}, correspondingly [28]. Figure 4.3 demonstrates the mechanism of the colorimetric detection for heavy metal ions using chemically functionalized AgNPs.

In aquatic water bodies, metal ions' detection is significant for ecosystem restoration, planning remediation strategies, and maintaining water quality. Organic molecules offer promising economically viable, robust, portable, and simple sensing systems for metal ions, plasmon-enhanced spectroscopy using plasmonic particles, and colorimetry decorated and tailored in point or field in care settings. A microwave-assisted method was used to fabricate AgNPs functionalized with pectin polymer (P-AgNPs) to detect heavy metals, such as Cr^{6+}, Cr^{3+}, Mn^{2+}, Hg^+, As^{5+}, Fe^{2+}, and Se^{4+} in an aqueous solution. A reddish-brown color was observed

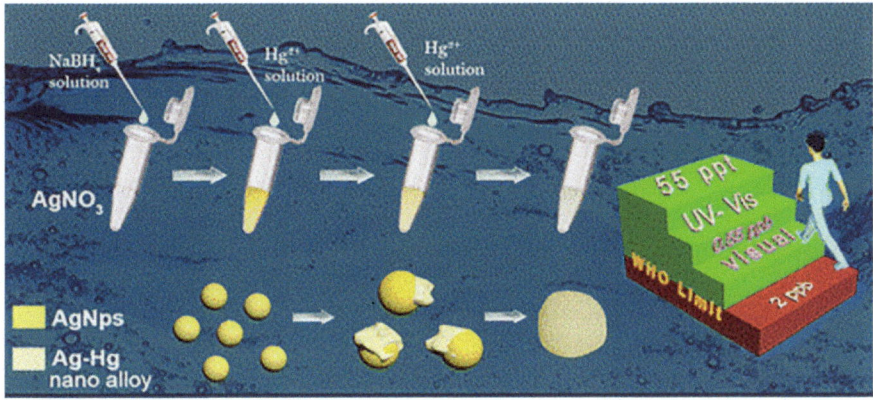

FIGURE 4.3 Schematic illustration of chemically synthesized AgNPs for sensing heavy metal ions in the water sample. (Reprinted with permission from Ref. [2]. Copyright 2014 American Chemical Society.)

when the as-synthesized P-AgNPs were added to water samples containing As^{5+} and Cr^{3+} ions, whereas Fe^{2+} and Mn^{2+} ions were brown and black appears to the water samples. The Raman measurements have also been recorded of diverse concentrations of Cr^{6+}, Cr^{3+}, Mn^{2+}, Se^{4+}, Hg^{+}, and As^{5+} ions introduced with developed colloidal P-AgNPs. An alteration occurs in the spectral features of the peaks of P-AgNPs, after the addition of metal ions to the P-AgNPs probe solution.

The formation of new characteristic bands occurs thru these P-AgNPs. As SERS signals, important improvement has been observed in the intensity of these bands. In an aqueous solution, these peculiar peaks can be utilized to detect heavy metal ions. In water samples, the detection of metal ion contaminants using the prompt, simple, portable, and easily accessible tool is demonstrated using P-AgNPs as a plasmon, enhancing spectroscopic and colorimetric assay [29]. Figure 4.4 shows the selectivity results in the presence of different metal ions of the detection probe.

In aqueous solutions, for detecting Cu^{2+} ions, humic acid-functionalized silver NPs (HA-AgNPs) were effectively developed and utilized. A polydispersity index of 0.447 and 101.4 nm of average hydrodynamic diameter showed by the

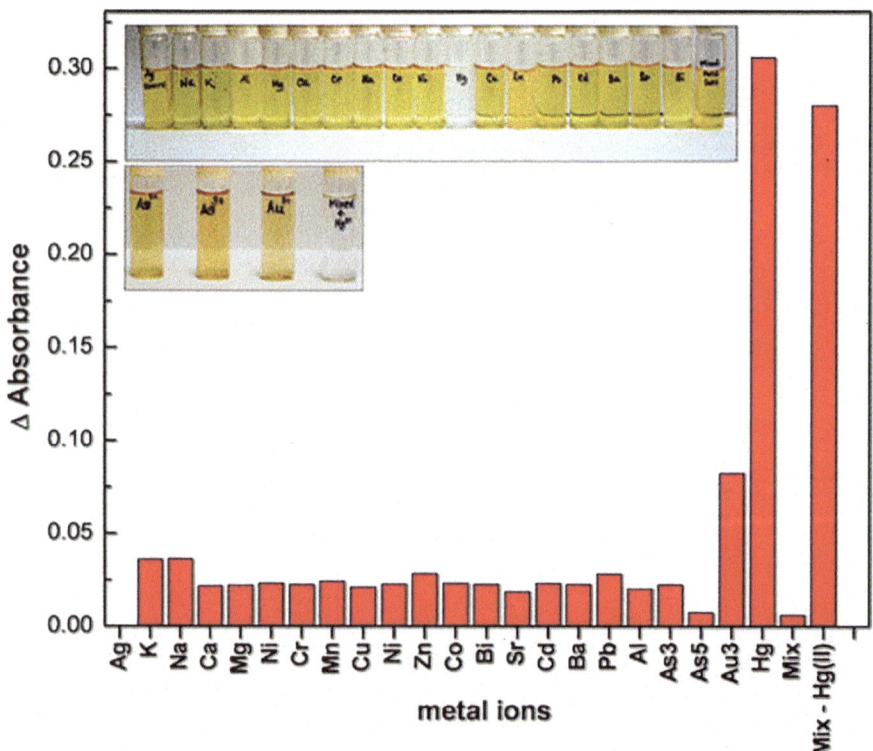

FIGURE 4.4 The selectivity studies using AgNPs in the existence of diverse metal ions. (Reprinted with permission from Ref. [2]. Copyright 2014 American Chemical Society.)

HA-AgNPs. The absorbance spectra of HA-AgNPs show the distinctive LSPR peak at 408.3 nm of synthesized AgNPs. Thru the corresponding optical absorbance spectra and change in their surface morphology, the aggregation of HA-AgNPs was demonstrated after adding Cu^{2+} ions. In the range of 0.00–1.25 mM of Cu^{2+} ion concentrations, the synthesized HA-AgNPs exhibited a strong linear response with a limit of quantification (LOQ) of 14.8094 ± 0.3636 mg L^{-1}, a limit of blank (LOB) of 0.1214 ± 0.0065 mg L^{-1}, and a LOD of 4.4428 ± 0.1091 mg L^{-1}. This calibration curve can be utilized to measure Cu^{2+} concentrations within a 95% confidence level by statistical analysis. Based on a selectivity study in mixed metal ions present in drinking water, HA-AgNPs were selective for detecting Cu^{2+} ions. Rapid quantification of Cu^{2+} ions was done using HA-AgNPs as a biodegradable colorimetric detection probe [30].

In numerous types of pollution, the toxic chemicals result in regular release into the environment. Over conventional analytical methods, the plasmonic sensors have precise benefits. For the detection of many analytes comprising metallic ions, chemical species, and enzymes, the utmost plasmonic candidates are AgNPs used in SPR based nanosensors. AgNP-based plasmonic nanosensors have removed many toxic metal ions, such as Cu^{2+}, Hg^{2+}, Pb^{2+}, Cd^{2+}, and Co^{2+}. It was endeavored to put extra emphasis on the sensing mechanisms, i.e., "dimensional/morphological change," "oxidation/ reduction," and "aggregation/ anti-aggregation" [31].

Mercury is the utmost toxic heavy metal among all the other heavy metals reached through industrial and mining wastewater in living beings and is considered the major ecological contaminant. The AgNPs functionalized with 3-(trimethoxysilyl) propyl methacrylate (TMPM) were used to detect Hg^{2+} ions in aqueous media. Various analytical techniques were used to characterize the as-synthesized NPs and detection processes such as particle size analyzer, UV–vis, emission-dispersive X-ray spectroscopy (EDS), TEM, and zeta sizer. The AgNPs exhibited λ_{max} at 404 nm by the analysis of UV–vis study. The particles' zeta potential was found at -34.32 ± 3 mV with an average diameter of 10 ± 1.4 nm. After detecting Hg^{2+} ions thru AgNPs, pale yellow to colorless color change was observed. Thru the measurements of surface charge and particle size of AgNPs, the peak's disappearance was observed in absorption spectra in UV–vis spectrophotometer confirmed the aggregation process after the addition of Hg^{2+} ions. The synthesized detection probe was further tested for the selectivity studies, which showed high selectivity toward Hg^{2+} ions in mixed metal ion solution. The gel-based and paper-based approaches were also established to detect the Hg^{2+} in an aqueous sample. A linear relationship of y confirmed the detection of mercury $=0.0007x-0.0081$ among the absorbance of AgNPs and the concentration of Hg^{2+} ion with a regression coefficient value $R^2=0.9816$. The atomic absorption spectrophotometric analysis was used to estimate mercury, which was found at 62.3 ± 1.9 nM. Toward the Hg^{2+} detection at the nanomolar level, the developed probe is exceedingly specific and responsive. Therefore, for effective estimation and detection of Hg^{2+} from many ecological samples, the developed probe is a prodigious tool [32].

Due to various health effects and the high toxicity of arsenic, a cost-effective, eco-friendly, and simple colorimetric method has been developed to select As^{3+} in an aqueous medium. The chemical reduction method was used to develop AgNPs functionalized with PEG fruitfully to detect heavy metal ions, such as Co^{2+}, As^{3+}, Cu^{2+}, Cd^{2+}, Pb^{2+}, Fe^{3+} Mg^{2+}, Hg^{2+}, and Ni^{2+} metal ions. Because of the aggregation of PEG functionalized AgNPs, a change in colors was observed from yellow to bluish after adding arsenic into the aqueous samples. Absorption spectra were measured through UV–vis of the arsenic samples to derive a detectable range of As^{3+} concentration. The detection probe's detection limit is 1 ppb the displayed by the results with outstanding linearity of ∼0.99 [33].

The Hg^{2+} ion detection has been done on AgNPs as a precise, sensitive colorimetric detection method. For preparing AgNPs, methylenediphosphonic acid (MDP) was used as a functionalizing agent using a chemical reduction method. Different analytical tools were used to characterize as-synthesized NPs, such as TEM, particle size analyzer, EDS, UV–vis spectroscopy, and zeta sizer. The NPs exhibited the zeta potential of −37.56±3 mV and spherical, which possess 8±1.1 nm value of hydrodynamic diameter. The synthesized detection probe displayed LOD for Hg^{2+} ions at a nanomolar level in the presence of antioxidant GSH. Because of the disappearance of the yellow color of the probe and its SPR with increasing Hg^{2+} concentrations, the sensing mechanism confirmed the aggregation process. The absorbance at 396 nm was observed, which showed excellent linearity between Hg^{2+} concentrations from 20 to 100 nM. The developed probe showed LOD at 8 nM. The optimizing parameters such as saline content, temperature, and different pH, not considerably affect the sensing capability of the probe. The probe selectively detects the Hg^{2+} ions in mixed metal ion solution. The amount of Hg^{2+} was estimated at 124 nM in a real water sample collected from Noyyal Orathuppalayam Reservoir and Dam using the probe. The developed probe showed excellent results toward quantitative and qualitative detection of Hg^{2+} from aqueous media as a colorimetric assay [34].

Gelatin functionalized AgNPs were developed by chemical synthesis and used for sensing Hg^{2+}. For the fabrication of AgNPs functionalized with gelatin, a chemical reduction method was utilized. The paper substrate, hydrogel network, and solution phases were used to carry out the colorimetric sensing of Hg^{2+} ions. Ag/Hg amalgam formation was due to the aggregation of AgNPs, which was further confirmed by the tuning of color from yellow to colorless with the addition of Hg^{2+} ions. The LOD was found 25 nM of as-synthesized nanosensor probe for the detection of Hg^{2+} ions.

3D hydrogel matrix and disposable paper strips were used to sense Hg^{2+} ions. It was under optimized conditions concerning selectivity and sensitivity for detecting Hg^{2+} ions in both cases; the sensor displayed similar properties. For many applications comprising antimicrobial coatings, sensing of heavy metals, food monitoring, and catalysis, these AgNPs functionalized by gelatin detection probes are exceedingly promising and versatile [35].

In a similar study, using starch as surface capping and a reducing agent, the silver NPs functionalized with starch were developed (AgNPST) with 2–10 nm of a size range. The zeta potential analysis, electron microscopy, and DLS were utilized to characterize the synthesized NPs for stability, hydrodynamic size, and shape. When excited at 350 nm, NPs showed fluorescence peaks of 400–680 nm and LSPR at 402 nm. The results showed AgNPST exhibited different SPR shifts when interacting with diverse heavy metals such as Ni^{2+}, Cu^{2+}, Pb^{2+}, Zn^{2+}, As^{5+}, and Hg^{2+}. However, in 1–10 ppm of concentration range, Hg^{2+} displayed the concentration-dependent visible yellow color besides maximum SPR shift. AgNPST could quantify and detect in vitro H_2O_2 besides cellular (*E. coli*) reactive oxygen species (ROS) confirmed by analyzing colorimetric assay, LSPR, and ROS by fluorescence. The results showed that to detect-free oxygen radicals and toxic metals such as Hg^{2+} ions, AgNPST may be used as a multifunctional sensor in H_2O_2 in vitro and biological milieu [36].

Pb^{2+}, Hg^{2+}, and Cd^{2+} metals ions were detected using AgNPs functionalized by N-(2-hydroxybenzyl)-isoleucine (ILP) and N-(2-hydroxybenzyl)-valine (VP) organic ligands as a surface functionalizing and reducing agent. The colorimetric metal ion selectivity and sensitivity of ILP- and VP-AgNPs were modified by integrating co-stabilizing agents. Significantly, in the polluted groundwater samples, likewise, the ILP functionalized AgNPs selectively recognized Cd^{2+} ions [37].

Wu et al. [38] and Ravindran et al. [39] reported cost-effective citrate-capped AgNPs for detecting Cr^{6+}/Cr^{3+} ions. The chelation of Cr^{3+} ions in both cases persuaded crosslinking aggregation of AgNPs with two carboxyls and one hydroxyl group of two citrate ions. Based on the alteration of LSPR scattering signals, bioimaging for Cr^{3+} in live cells was confirmed in the first case. The second case presented the lowest LOD in real samples to detect Cr^{6+} ions by colorimetric method using non-functionalized AgNPs. Currently, AgNPs are functionalized with AA developed by Wu et al. [40], where AA functions as a reducing and capping agent for Cr^{6+}, and a chelating agent for Cr^{3+}. The LOD reported by Ravindran et al. [41] showed better results, showing an LOD of 5×10^{-8} M for Cr^{6+} ions. Mn^{2+} ions were detected colorimetrically using AgNPs based on SPR [42]. They synthesized functionalized AgNPs using melamine (MA)-4- and mercaptobenzoic acid (4-MBA)- modified (AgNPs@4- MA-MBA). Currently, silver NPs functionalized with tripolyphosphate- (AgNPs@$P_3O_{10}5$-) were also synthesized for the detection of Mn^{2+} ions [43]. In both cases, because of the ion templated chelation in the presence of Mn^{2+}, the AgNPs@$P_3O_{10}5$ and AgNPs@4-MA-MBA- aggregated to form six-coordinated structures. In drinking water, the rapid determination of Mn^{2+} has been done using these methods. Though for the synthesis of AgNPs@4-MBA-MA, easily oxidized and unstable organic reagents were used. The advantages of using biocompatible inorganic, nontoxic and stable reagents have been presented thru the developed AgNPs@$P_3O_{10}5$ detection probe for Mn^{2+} ions.

4.5 CONCLUSIONS, FUTURE PERSPECTIVES, AND CHALLENGES

This chapter comprised the chemically synthesized silver and AuNPs as a colorimetric probe for metal ion sensing applications in aqueous media. The reported examples view the high creativity of researchers in developing functionalized NP-based selective and more efficient systems for environmental applications. The aims of the development of both the silver and gold nanosystems are similar, which comprised as the following manner:

- high sensitivity and selectivity,
- attain low limits of detection,
- detection without the necessity for sample pretreatment or preconcentration,
- rapid and simple determinations,
- real-time quantification and measurement, and
- low-cost methods that do not require costly instrumentation.

Despite their benefits, i.e., rapid, simple, portable, inexpensive, and easy-to-use, the colorimetric detection systems show a good detection limit compared to paper-based analytical devices. Instead, clinics, homes, resource-limited conditions, and area applications require nanosystems developed using non-technical, stable, biocompatible, and inexpensive inorganic reagents. They are simple, low-cost, practical, and facial approaches. Fluorescent methods present higher sensitivity than color systems based on emissive NPs, permitting faster qualitative and quantitative information. However, color-NP-based methods condensed the difficulty essential of conventional detection methods. More sophisticated nanosystems are desirable to detect different metal ions to improve their stability and multifunctionality.

REFERENCES

1. Chen, M. S., and D. W. Goodman. "The structure of catalytically active gold on titania." *Science* 306, no. 5694 (2004): 252–255. doi:10.1126/science.1102420.
2. Bhattacharjee, Yudhajit, and Amarnath Chakraborty. "Label-free cysteamine-capped silver nanoparticle-based colorimetric assay for Hg (II) detection in water with subnanomolar exactitude." *ACS Sustainable Chemistry & Engineering* 2, no. 9 (2014): 2149–2154. doi:10.1021/sc500339n.
3. Lim, Dong-Kwon, In-Jung Kim, and Jwa-Min Nam. "DNA-embedded Au/Ag core–shell nanoparticles." *Chemical Communications* 42 (2008): 5312–5314. doi:10.1039/B810195G.
4. Leesutthiphonchai, Wiphawee, Wijitar Dungchai, Weena Siangproh, Nattaya Ngamrojnavanich, and Orawon Chailapakul. "Selective determination of homocysteine levels in human plasma using a silver nanoparticle-based colorimetric assay." *Talanta* 85, no. 2 (2011): 870–876. doi:10.1016/j.talanta.2011.04.041.
5. Li, Haibing, Zhimin Cui, and Cuiping Han. "Glutathione-stabilized silver nanoparticles as colorimetric sensor for Ni^{2+} ion." *Sensors and Actuators B: Chemical* 143, no. 1 (2009): 87–92. doi:10.1016/j.snb.2009.09.013.
6. Liang, Aihui, Qingye Liu, Guiqing Wen, and Zhiliang Jiang. "The surface-plasmon-resonance effect of nanogold/silver and its analytical applications." *TrAC – Trends in Analytical Chemistry* 37 (2012): 32–47. doi:10.1016/j.trac.2012.03.015.

7. Murphy, Catherine J., Anand M. Gole, Simona E. Hunyadi, John W. Stone, Patrick N. Sisco, Alaaldin Alkilany, Brian E. Kinard, and Patrick Hankins. "Chemical sensing and imaging with metallic nanorods." *Chemical Communications* 5 (2008): 544–557. doi:10.1039/B711069C.

8. Obare, Sherine O., Rachel E. Hollowell, and Catherine J. Murphy. "Sensing strategy for lithium ion based on gold nanoparticles." *Langmuir* 18, no. 26 (2002): 10407–10410. doi:10.1021/la0260335.

9. Liu, Juewen, and Yi Lu. "Colorimetric biosensors based on DNAzyme-assembled gold nanoparticles." *Journal of Fluorescence* 14, no. 4 (2004): 343–354. doi:10.1023/B:JOFL.0000031816.06134.d3.

10. Pearson, Ralph G. "Recent advances in the concept of hard and soft acids and bases." *Journal of Chemical Education* 64, no. 7 (1987): 561. doi:10.1021/ed064p561.

11. Zhu, Lian, Lili Xu, Baozhen Huang, Ningming Jia, Liang Tan, and Shouzhuo Yao. "Simultaneous determination of Cd^{2+} and Pb^{2+} using square wave anodic stripping voltammetry at a gold nanoparticle-graphene-cysteine composite modified bismuth film electrode." *Electrochimica Acta* 115 (2014): 471–477. doi:10.1016/j.electacta.2013.10.209.

12. Ting, Siong Luong, Shu Jing Ee, Arundithi Ananthanarayanan, Kam Chew Leong, and Peng Chen. "Graphene quantum dots functionalized gold nanoparticles for sensitive electrochemical detection of heavy metal ions." *Electrochimica Acta* 172 (2015): 7–11. doi:10.1016/j.electacta.2015.01.026.

13. Shi, Lei, Yangyang Li, Xiaojiao Rong, Yan Wang, and Shiming Ding. "Facile fabrication of a novel 3D graphene framework/Bi nanoparticle film for ultrasensitive electrochemical assays of heavy metal ions." *Analytica Chimica Acta* 968 (2017): 21–29. doi:10.1016/j.aca.2017.03.013.

14. Sulistiawaty, Lilis, Sri Sugiarti, and Noviyan Darmawan. "Detection of Hg^{2+} metal ions using silver nanoparticles stabilized by gelatin and Tween-20." *Indonesian Journal of Chemistry* 15, no. 1 (2015): 1–8. doi:10.22146/ijc.21216.

15. Sarkar, Probir Kumar, Animesh Halder, Nabarun Polley, and Samir Kumar Pal. "Development of highly selective and efficient prototype sensor for potential application in environmental mercury pollution monitoring." *Water, Air, & Soil Pollution* 228, no. 8 (2017): 1–11. doi:10.1007/s11270-017-3479-1.

16. Vaid, Kalyan, Jasmeen Dhiman, Nikita Sarawagi, and Vanish Kumar. "Experimental and computational study on the selective interaction of functionalized gold nanoparticles with metal ions: Sensing prospects." *Langmuir* 36, no. 41 (2020): 12319–12326. doi:10.1021/acs.langmuir.0c02280.

17. Zhao, Chen, Guowei Zhong, Da-Eun Kim, Jinxia Liu, and Xinyu Liu. "A portable lab-on-a-chip system for gold-nanoparticle-based colorimetric detection of metal ions in water." *Biomicrofluidics* 8, no. 5 (2014): 052107. doi:10.1063/1.4894244.

18. Anand, Gopalan Sai, Anantha Iyengar Gopalan, Shin-Won Kang, and Kwang-Pill Lee. "Development of a surface plasmon assisted label-free calorimetric method for sensitive detection of mercury based on functionalized gold nanorods." *Journal of Analytical Atomic Spectrometry* 28, no. 4 (2013): 488–498. doi:10.1039/C3JA30300D.

19. Boruah, Bijoy Sankar, and Rajib Biswas. "Localized surface plasmon resonance based U-shaped optical fiber probe for the detection of Pb^{2+} in aqueous medium." *Sensors and Actuators B: Chemical* 276 (2018): 89–94. doi:10.1016/j.snb.2018.08.086.

20. Chai, Fang, Chungang Wang, Tingting Wang, Lu Li, and Zhongmin Su. "Colorimetric detection of Pb^{2+} using glutathione functionalized gold nanoparticles." *ACS Applied Materials & Interfaces* 2, no. 5 (2010): 1466–1470. doi:10.1021/am100107k.

21. Yuan, Huizhen, Wei Ji, Shuwen Chu, Qiang Liu, Siyu Qian, Jianye Guang, Jiabin Wang, Xiuyou Han, Jean-Francois Masson, and Wei Peng. "Mercaptopyridine-functionalized gold nanoparticles for fiber-optic surface plasmon resonance Hg^{2+} sensing." *ACS Sensors* 4, no. 3 (2019): 704–710. doi:10.1021/acssensors.8b01558.

22. Zhang, Yang, Ruixi Bai, Zhigang Zhao, Qiuxia Liao, Peng Chen, Wanghuan Guo, Chunqing Cai, and Fan Yang. "Highly selective and sensitive probes for the detection of Cr^{6+} in aqueous solutions using diglycolic acid-functionalized Au nanoparticles." *RSC Advances* 9, no. 19 (2019): 10958–10965. doi:10.1039/C9RA00010K.

23. Chen, Zhengbo, Junxia Guo, He Ma, Tong Zhou, and Xiaoxiao Li. "A simple colorimetric sensor for potassium ion based on DNA G-quadruplex conformation and salt-induced gold nanoparticles aggregation." *Analytical Methods* 6, no. 19 (2014): 8018–8021. doi:10.1039/C4AY01025F.

24. Kim, Sudeok, Juseon Kim, Na Hee Lee, Hyun Hye Jang, and Min Su Han. "A colorimetric selective sensing probe for calcium ions with tunable dynamic ranges using cytidine triphosphate stabilized gold nanoparticles." *Chemical Communications* 47, no. 37 (2011): 10299–10301. doi:10.1039/C1CC13489B.

25. Kim, Sunghyun, Jeong Won Park, Dongkyu Kim, Daejin Kim, In-Hyun Lee, and Sangyong Jon. "Bioinspired colorimetric detection of Ca^{2+} ions in serum using calsequestrin-functionalized gold nanoparticles." *Angewandte Chemie International Edition* 48, no. 23 (2009): 4138–4141. doi:10.1002/anie.200900071.

26. Zhang, Jia, Yong Wang, Xiaowen Xu, and Xiurong Yang. "Specifically colorimetric recognition of calcium, strontium, and barium ions using 2-mercaptosuccinic acid-functionalized gold nanoparticles and its use in reliable detection of calcium ion in water." *Analyst* 136, no. 19 (2011): 3865–3868. doi:10.1039/C1AN15175D.

27. Oliveira, Elisabete, Cristina Núñez, Hugo Miguel Santos, Javier Fernández-Lodeiro, Adrián Fernández-Lodeiro, José Luis Capelo, and Carlos Lodeiro. "Revisiting the use of gold and silver functionalised nanoparticles as colorimetric and fluorometric chemosensors for metal ions." *Sensors and Actuators B: Chemical* 212 (2015): 297–328. doi:10.1016/j.snb.2015.02.026.

28. Minh, Phung Nhat, Van-Tuan Hoang, Ngo Xuan Dinh, Ong Van Hoang, Nguyen Van Cuong, Dang Thi Bich Hop, Tran Quoc Tuan, Nguyen Tien Khi, Tran Quang Huy, and Anh-Tuan Le. "Reduced graphene oxide-wrapped silver nanoparticles for applications in ultrasensitive colorimetric detection of Cr^{6+} ions and the carbaryl pesticide." *New Journal of Chemistry* 44, no. 18 (2020): 7611–7620. doi:10.1039/D0NJ00947D.

29. Sharma, Sweta, Aarti Jaiswal, and K. N. Uttam. "Colorimetric and surface enhanced Raman scattering (SERS) detection of metal ions in aqueous medium using sensitive, robust and novel pectin functionalized silver nanoparticles." *Analytical Letters* 53, no. 15 (2020): 2355–2378. doi:10.1080/00032719.2020.1743715.

30. Lopez, Edgar Clyde R., Michael Angelo Zafra, Jon Nyner L. Gavan, Emil David A. Villena, Francis Eric P. Almaquer, and Jem Valerie D. Perez. "Humic acid functionalized-silver nanoparticles as nanosensor for colorimetric detection of Cu^{2+} ions in aqueous solutions." In *Key Engineering Materials*, vol. 831, pp. 142–150. Trans Tech Publications Ltd., 2020. doi:10.4028/www.scientific.net/KEM.831.142.

31. Amirjani, Amirmostafa, and Davoud Fatmehsari Haghshenas. "Ag nanostructures as the surface plasmon resonance (SPR)- based sensors: A mechanistic study with an emphasis on heavy metallic ions detection." *Sensors and Actuators B: Chemical* 273 (2018): 1768–1779. doi:10.1016/j.snb.2018.07.089.

32. Balasurya, S., Parvaiz Ahmad, Ajith Mesmin Thomas, Lija L. Raju, Arunava Das, and S. Sudheer Khan. "Rapid colorimetric and spectroscopy based sensing of mercury by surface functionalized silver nanoparticles in the presence of tyrosine." *Optics Communications* 464 (2020): 125512. doi:10.1016/j.optcom.2020.125512.

33. Boruah, Bijoy Sankar, Nikhil Kumar Daimari, and Rajib Biswas. "Functionalized silver nanoparticles as an effective medium towards trace determination of As^{3+} in aqueous solution." *Results in Physics* 12 (2019): 2061–2065. doi:10.1016/j.rinp.2019.02.044.

34. Janani, B., Asad Syed, Ajith M. Thomas, Najat Marraiki, Sarah Al-Rashed, Abdallah M. Elgorban, Lija L. Raju, Arunava Das, and S. Sudheer Khan. "Enhanced SPR signals based on methylenediphosphonic acid functionalized Ag NPs for the detection of Hg^{2+} in the presence of an antioxidant glutathione." *Journal of Molecular Liquids* 311 (2020): 113281. doi:10.1016/j.molliq.2020.113281.

35. Jeevika, Alagan, and Dhesingh Ravi Shankaran. "Functionalized silver nanoparticles probe for visual colorimetric sensing of mercury." *Materials Research Bulletin* 83 (2016): 48–55. doi:10.1016/j.materresbull.2016.05.029.

36. Ban, Deependra Kumar, and Subhankar Paul. "Rapid colorimetric and spectroscopy based sensing of heavy metal and cellular free oxygen radical by surface functionalized silver nanoparticles." *Applied Surface Science* 458 (2018): 245–251. doi:10.1016/j.apsusc.2018.07.069.

37. Kumar, V. Vinod, and Savarimuthu Philip Anthony. "Silver nanoparticles based selective colorimetric sensor for Cd^{2+}, Hg^{2+} and Pb^{2+} ions: Tuning sensitivity and selectivity using co-stabilizing agents." *Sensors and Actuators B: Chemical* 191 (2014): 31–36. doi:10.1016/j.snb.2013.09.089.

38. Wu, Tong, Chun Liu, Ke Jun Tan, Ping Hu, and Cheng Zhi Huang. "Highly selective light scattering imaging of Cr^{3+} in living cells with silver nanoparticles." *Analytical and Bioanalytical Chemistry* 397, no. 3 (2010): 1273–1279. doi:10.1007/s00216-010-3619-6.

39. Ravindran, Aswathy, M. Elavarasi, T. C. Prathna, Ashok M. Raichur, N. Chandrasekaran, and Amitava Mukherjee. "Selective colorimetric detection of nanomolar Cr^{6+} in aqueous solutions using unmodified silver nanoparticles." *Sensors and Actuators B: Chemical* 166 (2012): 365–371. doi:10.1016/j.snb.2012.02.073.

40. Wu, Xiaoyan, Yunbo Xu, Yangjun Dong, Xue Jiang, and Ningning Zhu. "Colorimetric determination of hexavalent chromium with ascorbic acid capped silver nanoparticles." *Analytical Methods* 5, no. 2 (2013): 560–565. doi:10.1039/C2AY25989C.

41. Wu, Yuangen, Le Liu, Shenshan Zhan, Faze Wang, and Pei Zhou. "Ultrasensitive aptamer biosensor for As^{3+} detection in aqueous solution based on surfactant-induced aggregation of gold nanoparticles." *Analyst* 137, no. 18 (2012): 4171–4178. doi:10.1039/C2AN35711A.

42. Zhou, Ying, Hong Zhao, Chang Li, Peng He, Wenbo Peng, Longfei Yuan, Lixi Zeng, and Yujian He. "Colorimetric detection of Mn^{2+} using silver nanoparticles cofunctionalized with 4-mercaptobenzoic acid and melamine as a probe." *Talanta* 97 (2012): 331–335. doi:10.1016/j.talanta.2012.04.041.

43. Gao, Yue-Xia, Jun-Wei Xin, Zhe-Yu Shen, Wei Pan, Xing Li, and Ai-Guo Wu. "A new rapid colorimetric detection method of Mn^{2+} based on tripolyphosphate modified silver nanoparticles." *Sensors and Actuators B: Chemical* 181 (2013): 288–293. doi:10.1016/j.snb.2013.01.079.

5 Paper-Based Plasmonic Nanosensors

5.1 INTRODUCTION

For printing, writing, packaging, and drawing, paper is an eminent substantial. From its physical properties beyond these traditional and simple main stems, the potential utility of paper is also considered as a potential sensor toward detecting various analytes. Its pulp processing can be made flexible, lightweight, and thin because it is a very sophisticated material. The cellulose fiber can be exceedingly fascinating for specific applications as the main constituent of paper. It allows liquid to enter its hydrophilic fiber matrix [1].

Through altering properties, such as reactivity, permeability, and hydrophilicity, cellulose fibers can be functionalized [2]. Currently, owing to its versatility, high abundance, and cost-effective availability, the paper has fascinated enormous consideration in clinical and analytical chemistry as a potential substance for devices and sensors [3–5]. Portability, flexibility, straightforward operation, and disposable nature are the properties integrated by these analytical devices. Glucose was detected in urine samples semi-quantitatively. The first paper device was demonstrated in 1956, being a well-known example [6], which was additionally advanced into immunochromatographic paper test strips with the pregnancy test kit better recognized as dipstick tests or lateral flow [7]. These immunoassays comprise a test line (immobilized on the surface to capture antibodies), a reagent pad (comprising antibodies precise to the target antigen, which are coupled to a signal indicator), and a sample pad paper strip (for introducing the sample). The sample migrates along the paper strip through capillary forces when introduced at the sample pad because of the antigen's existence in the sample that binds to the signal antibody.

The developed antigen-antibody/signal endures flowing together with the paper strip where the capture antibody is successively captured at the surface to show a positive result. The signal indicator is characteristically AuNPs or colored latex microspheres [8]. Although consistent with showing a qualitative "yes/no" type of detection, these low-cost and simple devices are usually imperfect. A move in concentration from elementary design concepts to further innovative patterning and manufacture methods has been seen in the last few years to attain additional quantitative and correct results. For the detection of diverse analytes, the idea of fabricating microfluidic channels on paper (μPADs) was demonstrated by Whitesides and co-workers [9].

In contrast to the concentration of the analyte, the detection technique measures the color intensity of the analyte, and it is based on colorimetry.

DOI: 10.1201/9781003128281-5

Many innovative areas of exploration and fabrication have opened in microfluidic separation devices and paper-cut microfluidic devices [10–13]. As the solution travels up the paper, chromatographic separation of mixtures occurs [14–16]. Chemiluminescence (CL) [5], electrochemical [17–19], electrical [20,21], and electrochemiluminescence [22] are the techniques other than colorimetry [9,23–27] to detect analytes. These novel research avenues in sensors have attained. In terms of cost-effectiveness, simplicity, and sensitivity, these methods have their benefits and limitations. For lab-on-a-chip devices, the paper has developed a promising platform in the past few years owing to the development of paper-based microfluidics, in which complicated and large-scale laboratory tests could be executed.

In many applications where practical and simple analytical devices are exceedingly required, onsite, real-time, portable detection is vital, for example, in the environmental, food, and clinical areas. A challenge to gain additional rapid test results, which can be encountered through paper-based sensors, is a growing requirement for POC diagnostics with the spiraling costs of health care. This chapter focuses on the application areas for these sensors, existing analysis, fabrication methods, and the paper used for sensing methods. The future outlooks and present challenges will be discussed that are essential to be addressed to grasp their full potential for paper-based sensors.

5.2 FABRICATION OF PAPER-BASED SENSORS

5.2.1 Choices for Paper

Though the choice is based primarily on the precise application area and the developing steps essential in manufacturing a device, various paper materials are accessible. In recent years, for fabricating paper-based sensors, filter paper has seen extensive usage owing to its wicking ability in the microfluidic technologies and development of sensors [28,29]. Specifically, existence flow rate, particle retention, and porosity, the Whatman cellulose range distinguishing the types of filter paper is widespread having significant parameters. Whatman filter paper No. 1, with a medium flow rate and retention ability, has been utilized by various research groups, a standard grade filter paper [30–34]. Li et al. [29] used Whatman No. 4 filter paper to increase liquid penetration through coating it with a cellulose hydrophilization agent as a base for etching printing of hydrophilic channels. The cellulose fibers' swelling occurs through the solvent, which could hinder liquid penetration by restricting the capillary pores. Therefore, this type of filter paper has a larger pore size than the standard grade and is chosen as a sensor material. The diverse paper modifications or sorts of paper have been discovered because the Whatman filter paper does not always own the required physical characteristics. However, filter paper is extensively utilized in this field. Hydrophobic nitrocellulose membranes are used to immobilize enzymes [26], proteins [11], and DNA [35] because of their high tendency to non-specific binding to biomolecules. For a colorimetric assay, Lu et al. [36,37] synthesized a paper-based sensor using nitrocellulose membrane as the substrate. The deposition of an enzyme has been done by forming a wax barrier on the membrane through heating and printing. Though the wax penetration is slow compared to filter paper, while nitrocellulose membranes have a sensibly even pore size (0.45 μm), these are smooth, resulting in a more stable liquid flow and reproducibility paper. The usage of chemically changed cellulose fibers is another boulevard for exploration. Composite papers containing polyester and cellulose and ion-exchange cellulose papers are commercially accessible [19]. To produce paper-based sensing methods, they have reported other sorts of paper as an appropriate material, for example, glossy paper, rather than using filter paper as the key material. Glossy paper is an inorganic filler blended cellulose fiber with a flexible substrate. For the detection of ethanol, Arena et al. [20] used multi-walled carbon nanotubes (MWCNTs) for developing a flexible paper-based sensor as electrodes and indium tin oxide (ITO) nanoparticulate powder as a sensing material on glossy paper. For filter paper, glossy paper is a virtuous substitute owing to its relatively smooth surface and non-degradability. The fiber matrix is essential, specifically, when adjusting nanomaterials onto a surface.

5.2.2 PATTERNING AND FABRICATION

The choice of materials and techniques to develop paper devices are effective manufacture progression, simplicity, and cost-effectiveness. Many processes and techniques can be utilized to alter the goods of the paper, including physical deposition and chemical modification. They turn out to be available for direct use in a variety of applications or further modification [38], analogue plotting [25], photolithography [23,24], etching [23,27,29], and inkjet printing [39], paper cutting [10,11], plasma treatment [40,41], flexography printing [42], wax printing [43–45], laser treatment [46], and screen printing [4] are the methods used for the development of paper-based sensors. Contingent on the modification needed, and the material used, the selection of methods has been made. Paper-based microfluidics are focused on confining the liquid to a specific region on the paper. Therefore, to build up the active sensing element, first, we discuss some of these approaches followed by some other methods.

5.3 QUANTITATIVE ANALYSIS

Paper-based sensors offer the possibility to the users to develop mass-scalable and simple devices at an affordable cost. Transduction approaches are essential to produce an analytical device, which can cause additional instrumentation, materials, and reagents at additional complexity and cost. Low power methods, such as electrochemical and optical techniques, are best suitable as transducers to preserve portability, affordability, and simplicity. For quantitative analysis on paper, electrochemical [4,16,47,48], colorimetric [49], electrical conductivity [21], electrochemiluminescence [50], and CL [51] are the five utmost usually described techniques.

5.3.1 COLORIMETRIC DETECTION

Silver nanoparticles (AgNPs) have been used as paper devices by Ratnarathorn et al. [49] to detect Cu^{2+} ions colorimetrically. In the test zone, by the immobilization of AgNPs functionalized with dithiothreitol (DTT) and homocysteine (Hcy), the paper-based sensors were fabricated. The liquid was transported to the test zone after the development of the sensor by dropping the copper solution onto the loading zone. By binding Cu^{2+} to the amino and carboxyl functional groups on DTT and Hcy, an alteration in color was perceived due to the aggregation of Hcy-DTT-AgNPs induced by copper ions. The sensor showed the LOD up to 7.8 nM by the naked eye. Figure 5.1 shows the paper-based detection systems for arsenic detection in water samples.

AuNPs have been fabricated for the colorimetric detection of Cr^{3+} in water samples with no surface modification. Due to the aggregation of citrate-capped AuNPs after adding Cr^{3+} ions, a red shift in the UV–visible absorption maxima occurs from 526 to 714 nm. By the DLS measurements and SEM analysis, the aggregation of the AuNPs has been confirmed. Under the optimized experimental conditions at 714 nm to that at 526 nm (A714/526), a linear relationship

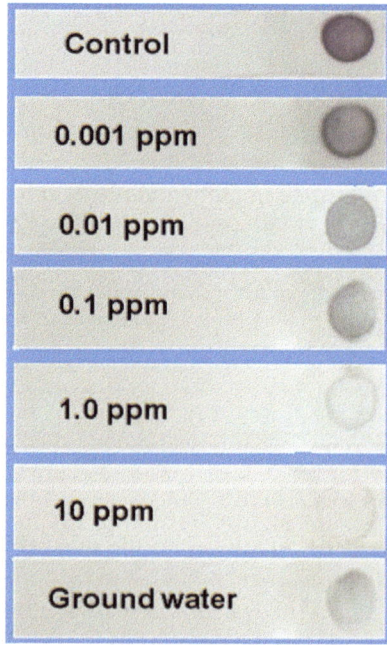

FIGURE 5.1 Analysis of arsenic with groundwater and diverse arsenic ($As^{3+}+As^{5+}$ 1:1) concentrations on a paper strip. Colorimetric response in the presence of arsenic-contaminated real groundwater sample (Right) while the color of the GNR-PEG-DMSA control shown (Extreme left). (Reprinted with permission from Ref. [57]. Copyright 2018 American Chemical Society.)

(correlation coefficient $r=0.997$) was established between the ratio of the absorbance and over the concentrations range of 10^{-3} to 10^{-6} M of Cr^{3+} ions. The developed detection probe showed virtuous selectivity toward Cr^{3+} ions with 1.06×10^{-7} M of the LOD. The method's potential was further confirmed by immobilization of AuNPs on Whatman filter paper strips and interacted with Cr^{3+} ions to compare the measured color intensity, which was calibrated in contrast to the concentration of the analyte. The selectivity of the paper strip-based device was successfully tested in the existence of an artificial mixture of interfering metal ions, which showed high selectivity toward Cr^{3+} ions. The as-synthesized AuNPs showed utmost importance as a paper strip-based detection probe for the onsite monitoring of successful Cr^{3+} ions in contaminated sites [52].

In a similar study, using virgin nanoporous silicon (NPSi) on a "Test Paper," an innovative paper-based sensing method has been developed, displaying excellent fluorescence stability and intense visible emission. For the detection of Cu^{2+} at μmol L^{-1} level, the visual fluorescence quenching "Test Paper" showed high selectivity and sensitivity. In the concentration range of $5 \times 10^{-7} \sim 50 \times 10^{-7}$ mol L^{-1}, the quantitative detection of Cu^{2+} ions has been done with the linear regression coefficient of $R^2=0.99$. The XPS analysis showed the fluorescence quenching mechanism of NPSi prober toward Cu^{2+} ions thru studying the change of surface chemistry of metal ions immersed NPSi and NPSi sensing probe. The outcomes showed that introducing a nonradiative recombination center and the oxidization state are responsible for the PL quenching and specify that SiH_x species noticeably donate to the PL emission of NPSi. By investigating this study, it was predicted that it shows significant importance to encourage the development of simple instruments for real-time, visible, and rapid detection of many toxicants [53].

A paper-based analytical device has been utilized to sense Hg^{2+} ions in water samples using AuNPs as an on-field colorimetric detection technique. The detection of Hg^{2+} ions is done by oligonucleotide sequences attached to unmodified AuNPs using thymine–Hg^{2+}– thymine (T–Hg^{2+}–T) coordination chemistry. The developed method showed label-free detection without time-consuming and complicated preparation processes of the thiolated probe. The developed probe could detect another metal ion in a mixture of metal ions through the oligonucleotide orders, introducing diverse degrees of aggregation of AuNPs. The alteration of various recognition outcomes was concentrated and transferred analytical devices based on cellulose paper to diminish the requirement of power for data transmission and analysis and remove the usage of sophisticated instruments.

In addition, using cloud computing through a smartphone, the data were then transmitted for the storage and readout of outcomes. The results showed a LOD for Hg^{2+} ion in river water and spiked pond up to 50 nM. The outcomes demonstrate that in resource-constrained locations, the developed method owns the competence for high-throughput and sensitive onsite detection of Hg^{2+} ions [54].

In another study, paper-based analytical devices (PADs) have been established to detect Hg^{2+} ion concentration using curcumin NPs (CURN). The wax dipping method was used to develop the paper-based probe by adding CURN as the sensing reagent. The analytical signal was defined as each test zone's mean color intensity, which was amplified on augmenting Hg^{2+} ion concentration. Various optimizing parameters, such as buffer, pH, the concentration of analyte and CURN, and ionic strength, were investigated to establish the optimum conditions. Without preconcentration, the LOD was found 0.17 µg mL^{-1} in the linear range of 0.5–20 µg mL^{-1} of Hg^{2+} ions, and after 50 times preconcentration, the LOD was found 0.003 µg mL^{-1}, and the calibration curve was found to be linear in the range of 0.01–0.4 µg mL^{-1} of Hg^{2+}. For ten replicate measurements, the relative standard deviation (RSD) of Hg^{2+} was 4.47% found up to 2 µg mL^{-1}. This chemosensor proved a potential method at different concentrations of Hg^{2+} in spiked water samples to analyze various satisfactory recoveries in different water sources [55].

For sensing Pb^{2+} and Cu^{2+} ions at their very low concentrations, Au-TA-DNS was developed as a smart gold nanosensor. In addition, of the metal ions to the sensor probe, the nanosensor shows visible blue color on paper strips due to the formation of NP aggregates and binding with metal ions. Au-TA-DNS also displays significant fluorescence quenching proportional to the concentration of ions due to dansyl fluorophore and binding with Cu^{2+} and Pb^{2+} in an aqueous medium. The fluorometric and colorimetric detection systems show ≤10.0 ppb LOD and are very selective for Cu^{2+} and Pb^{2+} ions. For field-test applications at a broad scale, for example, the monitoring process of water quality, the paper-based sensing method would be useful and has the advantage of low cost [56].

The GNR functionalized with dimercaptosuccinic acid was used to detect arsenic species by paper strip. The developed GNR-PEG-DMSA could quickly sense arsenic species in aqueous systems at deficient concentrations of arsenic pollution. The stepwise chemical conjugations of poly(ethyleneglycol) methyl ether thiol (mPEG-SH) with GNR followed by meso-2,3-dimercaptosuccinic acid (DMSA) were used to develop the sensor probe. On a paper substrate and in solution, the GNR-PEG-DMSA displays a visible change in color in the presence of both inorganic forms of arsenic: As^{5+} (arsenate) and As^{3+} (arsenite) ions. This is done due to an arsenic complex formation method thru the ion-induced aggregation of nanorods, which is comparative to different concentrations of arsenic. Spectroscopic and colorimetric investigates are selective, with a detection limit of ~1.0 ppb for both As^{5+} and As^{3+} ions to a great extent. In groundwater samples, for quantitative estimation of total arsenic, the GNR-PEG-DMSA displays outstanding potential. Under field-test mode in arsenic-level screening, the paper-based detection technique determines its utility [57].

In tap water, for the detection of Cd^{2+} ions, a monoclonal antibody (3A9) was developed based on a highly sensitive and novel immunochromatographic strip. The 50% inhibition concentration was 0.45 ng mL^{-1} of the antibody, which detect not metal-free EDTA and ITCBE and Cd(II)–1-(4-isothiocyanobenzyl) ethylenediamine-N,N,N′,N′-tetraacetic acid (EDTA) (Cd(II)–ITCBE) also showed no cross-reactivity with other heavy metal ions. The lower LOD was found 0.2 ng mL^{-1} by using a scanning reader for quantitative detection. For semiquantitative detection of the strip, the cutoff value was 5 ng mL^{-1}. The percent recovery in tap

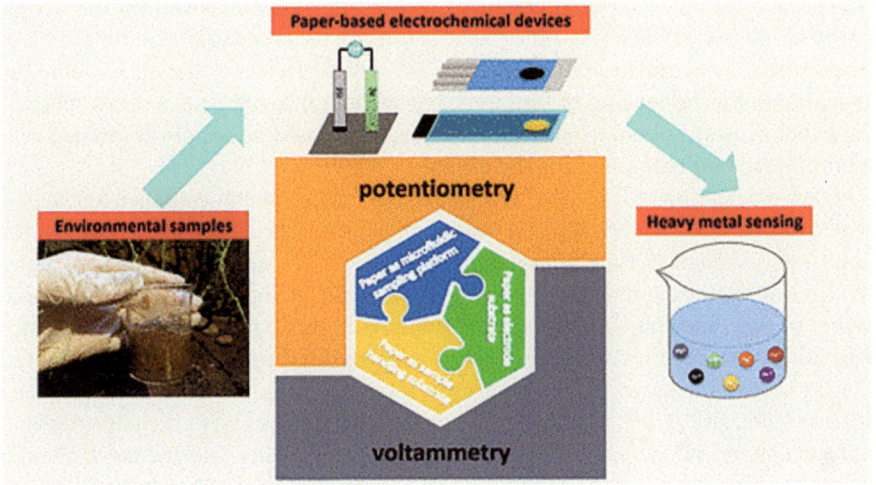

FIGURE 5.2 Schematic presentation of electrochemical recognition of heavy metal ions using paper-based detection. (Reprinted with permission from Ref. [4]. Copyright 2021 American Chemical Society.)

water samples ranged from 107.6% to 132% [58]. The schematic presentation of electrochemical detection of heavy metal ions is shown in Figure 5.2 using paper-based detection systems.

In industrial wastewater solutions, for the fast detection of Au^{3+} in the presence of Fe^{3+} ion, which is a common interfering ion, Apilux et al. [28] developed a novel lab-on-paper mutual with colorimetric and electrochemical sensing. The ICP-AES analysis demonstrated that the developed lab-on-paper showed the LOD up to can detect up to 1 ppm, which is a lower level of Au^{3+} ion concentration. With a correlation coefficient of 0.997, the as-synthesized sensor shows good linearity in the range of 1–200 ppm. The developed paper-based microfluidic pads showed multimodal detection as an advantage quickly, and it can also dispose of after usage.

AuNPs have been used as an on-field paper-based technique for sensing Hg^{2+} ions in water sources [59]. A simple image-processing approach and smartphone were used to analyze the many ensuing alterations in color, concentrated and transferred on paper-based devices. Using paper microfluidic pads, the need for an expensive labeled probe preparation process is diminished. The detection probe showed 50 nM of LOD for river and pond water.

Zhang et al. [60] developed a self-powered sensor for the sensing of Hg^{2+} ions. For the detection of Hg^{2+} ions, carbon nanotube (CNT) functionalized with Pt and AuNP (Pt/CNT), which serve as cathodic and anodic substrates used as changed paper electrodes. This developed probe shows that with a core component free of a mediator, the hollow channels are used to transport fluids. The low-cost sensor offers the complete benefit of POC testing and can detect at the picomolar level of Hg^{2+} ion concentrations.

The competence of electrochemical sensing devices improved by the use of paper-based micro-pads [19]. The paper acts as an excellent platform for the electrochemical detection system and assists as a good matrix for the measurements. In addition, for the sensing of Pb^{2+} ions, Nie et al. [19] developed an electrochemical paper sensor. For environmental analysis, this could be simply combined into a portable reader that advances the choice of the sensor.

For the detection of Co^{2+}, Hg^{2+}, Zn^{2+}, and Ag^{2+} heavy metal cations from wastewater, Feng et al. [61] developed a fluorometric paper-based sensor. Likewise, for colorimetric sensing of metal pollutants in water, a 3D paper-based vertical assay device was synthesized, allowing four tests in parallel to be carried out [62]. Three layers of tape and four layers of paper comprised this device, and a wax printer was utilized to generate the hydrophobic patterns. Hydrophilic regions permitted the liquid to flow through, which are the sample testing zones. The metal ions such as Cd^{2+}, Ni^{2+}, Cu^{2+}, and Cr^{3+} ions can be detected using specific chromogenic reagents stored in each sampling zone. The color intensity and the color change were subsequently quantified and imaged using a mobile phone camera.

5.4 CHALLENGES IN PAPER-BASED SENSORS AND FUTURE OUTLOOKS

Recent advances in developments and analysis techniques are still required to match the performance of conventional analytical methods. However, there is enormous potential in the paper as a platform for LOC devices. The often-encountered limitation is the overall sensitivity of the device. Paper-based electrochemical sensors (PESs) used in various application areas are novel alternative analytical methods for developing low-cost, simple, portable, disposable, and user-friendly sensing devices. In various areas such as clinical diagnostics, food, and environmental fields, PESs have been extensively used since 2009, where practical and straightforward analytical techniques are exceedingly required. Therefore, PESs are probably progressively developed in the future in contrast to other intricate measurable sensing techniques. In laboratories to commercial devices, for transformation from proof-of-concept strategies, potential utilities and recent advancements of PESs hold great promise. In the comprehensive performance of PESs as the major advances, development methods and selection of electrode materials show a vital part. For more functional and efficient platforms, multiplexed designs in origami, 3D, or 2D PAD configurations have also shown where complicated or multiple examinations could be achieved in a facile manner. However, to verify the real-life applicability and versatility of PESs, future developments need to be focused on.

- progress of diverse sensing techniques;
- modification and fabrication methods of electrode materials and paper substrate;
- emerging low-cost and simple mass production approaches, which will grasp fruitful commercialization;

- fabrication of automated and integrated devices, which have further purposes in simple techniques;
- To spread their benefits in remote areas, self-powered devices should also be considered.

To produce more stable, better devices that are proficient in detecting various environmental pollutants at high sensitivity and overcoming the above-discussed challenges, additional research must be focused on integrating functional materials onto the surface and development methods. It will be vital not to sacrifice the advantages of cost efficiency and simplicity of paper-based sensors to achieve these types of sensors.

REFERENCES

1. Martinez, Andres W., Scott T. Phillips, George M. Whitesides, and Emanuel Carrilho. "Diagnostics for the developing world: Microfluidic paper-based analytical devices." (2010): 3–10. doi:10.1021/ac9013989.
2. Bracher, Paul J., Malancha Gupta, and George M. Whitesides. "Patterning precipitates of reactions in paper." *Journal of Materials Chemistry* 20, no. 24 (2010): 5117–5122. doi:10.1039/C000358A.
3. Xu, Yuanhong, Mengli Liu, Na Kong, and Jingquan Liu. "Lab-on-paper micro- and nano-analytical devices: Fabrication, modification, detection and emerging applications." *Microchimica Acta* 183, no. 5 (2016): 1521–1542. doi:10.1007/s00604-016-1841-4.
4. Ding, Ruiyu, Yi Heng Cheong, Ashiq Ahamed, and Grzegorz Lisak. "Heavy metals detection with paper-based electrochemical sensors." (2021): 1880–1888. doi:10.1021/acs.analchem.0c04247.
5. Yu, Jinghua, Lei Ge, Jiadong Huang, Shoumei Wang, and Shenguang Ge. "Microfluidic paper-based chemiluminescence biosensor for simultaneous determination of glucose and uric acid." *Lab on a Chip* 11, no. 7 (2011): 1286–1291. doi:10.1039/C0LC00524J.
6. Comer, J. P. "Semiquantitative specific test paper for glucose in urine." *Analytical Chemistry* 28, no. 11 (1956): 1748–1750. doi:10.1021/ac60119a030.
7. Von Lode, Piia. "Point-of-care immunotesting: Approaching the analytical performance of central laboratory methods." *Clinical Biochemistry* 38, no. 7 (2005): 591–606. doi:10.1016/j.clinbiochem.2005.03.008.
8. Zhao, Weian, M. Monsur Ali, Sergio D. Aguirre, Michael A. Brook, and Yingfu Li. "Paper-based bioassays using gold nanoparticle colorimetric probes." *Analytical Chemistry* 80, no. 22 (2008): 8431–8437. doi:10.1021/ac801008q.
9. Martinez, Andres W., Scott T. Phillips, Manish J. Butte, and George M. Whitesides. "Patterned paper as a platform for inexpensive, low-volume, portable bioassays." *Angewandte Chemie* 119, no. 8 (2007): 1340–1342. doi:10.1002/ange.200603817.
10. Wang, Wei, Wen-Ya Wu, and Jun-Jie Zhu. "Tree-shaped paper strip for semi-quantitative colorimetric detection of protein with self-calibration." *Journal of Chromatography A* 1217, no. 24 (2010): 3896–3899. doi:10.1016/j.chroma.2010.04.017.
11. Fenton, Erin M., Monica R. Mascarenas, Gabriel P. López, and Scott S. Sibbett. "Multiplex lateral-flow test strips fabricated by two-dimensional shaping." *ACS Applied Materials & Interfaces* 1, no. 1 (2009): 124–129. doi:10.1021/am800043z.

12. Kauffman, Peter, Elain Fu, Barry Lutz, and Paul Yager. "Visualization and measurement of flow in two-dimensional paper networks." *Lab on a Chip* 10, no. 19 (2010): 2614–2617. doi:10.1039/C004766J.

13. Fu, Elain, Barry Lutz, Peter Kauffman, and Paul Yager. "Controlled reagent transport in disposable 2D paper networks." *Lab on a Chip* 10, no. 7 (2010): 918–920. doi:10.1039/B919614E.

14. Carvalhal, Rafaela Fernanda, Marta Simão Kfouri, Maria Helena de Oliveira Piazzetta, Angelo Luiz Gobbi, and Lauro Tatsuo Kubota. "Electrochemical detection in a paper-based separation device." *Analytical Chemistry* 82, no. 3 (2010): 1162–1165. doi:10.1021/ac902647r.

15. Carvalhal, Rafaela Fernanda, Emanuel Carrilho, and Lauro Tatsuo Kubota. "The potential and application of microfluidic paper-based separation devices." *Bioanalysis* 2, no. 10 (2010): 1663–1665. doi:10.4155/bio.10.138.

16. Shiroma, Leandro Yoshio, Murilo Santhiago, Angelo L. Gobbi, and Lauro T. Kubota. "Separation and electrochemical detection of paracetamol and 4-aminophenol in a paper-based microfluidic device." *Analytica Chimica Acta* 725 (2012): 44–50. doi:10.1016/j.aca.2012.03.011.

17. Dungchai, Wijitar, Orawon Chailapakul, and Charles S. Henry. "Electrochemical detection for paper-based microfluidics." *Analytical Chemistry* 81, no. 14 (2009): 5821–5826. doi:10.1021/ac9007573.

18. Nie, Zhihong, Frédérique Deiss, Xinyu Liu, Ozge Akbulut, and George M. Whitesides. "Integration of paper-based microfluidic devices with commercial electrochemical readers." *Lab on a Chip* 10, no. 22 (2010): 3163–3169. doi:10.1039/C0LC00237B.

19. Nie, Zhihong, Christian A. Nijhuis, Jinlong Gong, Xin Chen, Alexander Kumachev, Andres W. Martinez, Max Narovlyansky, and George M. Whitesides. "Electrochemical sensing in paper-based microfluidic devices." *Lab on a Chip* 10, no. 4 (2010): 477–483. doi:10.1039/B917150A.

20. Arena, A., N. Donato, G. Saitta, A. Bonavita, G. Rizzo, and G. Neri. "Flexible ethanol sensors on glossy paper substrates operating at room temperature." *Sensors and Actuators B: Chemical* 145, no. 1 (2010): 488–494. doi:10.1016/j.snb.2009.12.053.

21. Steffens, C., A. Manzoli, E. Francheschi, M. L. Corazza, F. C. Corazza, J. Vladimir Oliveira, and P. S. P. Herrmann. "Low-cost sensors developed on paper by line patterning with graphite and polyaniline coating with supercritical CO_2." *Synthetic Metals* 159, no. 21–22 (2009): 2329–2332. doi:10.1016/j.synthmet.2009.08.045.

22. Delaney, Jacqui L., Conor F. Hogan, Junfei Tian, and Wei Shen. "Electrogenerated chemiluminescence detection in paper-based microfluidic sensors." *Analytical Chemistry* 83, no. 4 (2011): 1300–1306. doi:10.1021/ac102392t.

23. Martinez, Andres W., Scott T. Phillips, and George M. Whitesides. "Three-dimensional microfluidic devices fabricated in layered paper and tape." *Proceedings of the National Academy of Sciences* 105, no. 50 (2008): 19606–19611. doi:10.1073/pnas.0810903105.

24. Klasner, Scott A., Alexander K. Price, Kurt W. Hoeman, Rashaun S. Wilson, Kayla J. Bell, and Christopher T. Culbertson. "Paper-based microfluidic devices for analysis of clinically relevant analytes present in urine and saliva." *Analytical and Bioanalytical Chemistry* 397, no. 5 (2010): 1821–1829. doi:10.1007/s00216-010-3718-4.

25. Bruzewicz, Derek A., Meital Reches, and George M. Whitesides. "Low-cost printing of poly (dimethylsiloxane) barriers to define microchannels in paper." *Analytical Chemistry* 80, no. 9 (2008): 3387–3392. doi:10.1021/ac702605a.

26. Martinez, Andres W., Scott T. Phillips, Emanuel Carrilho, Samuel W. Thomas III, Hayat Sindi, and George M. Whitesides. "Simple telemedicine for developing regions: Camera phones and paper-based microfluidic devices for real-time, off-site diagnosis." *Analytical Chemistry* 80, no. 10 (2008): 3699–3707. doi:10.1021/ac800112r.

27. Abe, Koji, Koji Suzuki, and Daniel Citterio. "Inkjet-printed microfluidic multiana-lyte chemical sensing paper." *Analytical Chemistry* 80, no. 18 (2008): 6928–6934. doi:10.1021/ac800604v.

28. Apilux, Amara, Wijitar Dungchai, Weena Siangproh, Narong Praphairaksit, Charles S. Henry, and Orawon Chailapakul. "Lab-on-paper with dual electrochemical/col-orimetric detection for simultaneous determination of gold and iron." *Analytical Chemistry* 82, no. 5 (2010): 1727–1732. doi:10.1021/ac9022555.

29. Li, Xu, Junfei Tian, Gil Garnier, and Wei Shen. "Fabrication of paper-based micro-fluidic sensors by printing." *Colloids and Surfaces B: Biointerfaces* 76, no. 2 (2010): 564–570. doi:10.1016/j.colsurfb.2009.12.023.

30. Ellerbee, Audrey K., Scott T. Phillips, Adam C. Siegel, Katherine A. Mirica, Andres W. Martinez, Pierre Striehl, Nina Jain, Mara Prentiss, and George M. Whitesides. "Quantifying colorimetric assays in paper-based microfluidic devices by measuring the transmission of light through paper." *Analytical Chemistry* 81, no. 20 (2009): 8447–8452. doi:10.1021/ac901307q.

31. Martinez, Andres W., Scott T. Phillips, Zhihong Nie, Chao-Min Cheng, Emanuel Carrilho, Benjamin J. Wiley, and George M. Whitesides. "Programmable diagnos-tic devices made from paper and tape." *Lab on a Chip* 10, no. 19 (2010): 2499–2504. doi:10.1039/C0LC00021C.

32. Bracher, Paul J., Malancha Gupta, and George M. Whitesides. "Patterned paper as a template for the delivery of reactants in the fabrication of planar materials." *Soft Matter* 6, no. 18 (2010): 4303–4309. doi:10.1039/C0SM00031K.

33. Songjaroen, Temsiri, Wijitar Dungchai, Orawon Chailapakul, and Wanida Laiwattanapaisal. "Novel, simple and low-cost alternative method for fabrication of paper-based microfluidics by wax dipping." *Talanta* 85, no. 5 (2011): 2587–2593. doi:10.1016/j.talanta.2011.08.024.

34. Luckham, Roger E., and John D. Brennan. "Bioactive paper dipstick sensors for acetylcholinesterase inhibitors based on sol–gel/enzyme/gold nanoparticle compos-ites." *Analyst* 135, no. 8 (2010): 2028–2035. doi:10.1039/C0AN00283F.

35. Cretich, Marina, Valentina Sedini, Francesco Damin, Maria Pelliccia, Laura Sola, and Marcella Chiari. "Coating of nitrocellulose for colorimetric DNA microarrays." *Analytical Biochemistry* 397, no. 1 (2010): 84–88. doi:10.1016/j.ab.2009.09.050.

36. Lu, Yao, Weiwei Shi, Jianhua Qin, and Bingcheng Lin. "Fabrication and char-acterization of paper-based microfluidics prepared in nitrocellulose membrane by wax printing." *Analytical Chemistry* 82, no. 1 (2010): 329–335. doi:10.1021/ac9020193.

37. Lu, Yao, Bingcheng Lin, and Jianhua Qin. "Patterned paper as a low-cost, flex-ible substrate for rapid prototyping of PDMS microdevices via "liquid molding"." *Analytical Chemistry* 83, no. 5 (2011): 1830–1835. doi:10.1021/ac102577n.

38. Li, Xu, David R. Ballerini, and Wei Shen. "A perspective on paper-based microflu-idics: Current status and future trends." *Biomicrofluidics* 6, no. 1 (2012): 011301. doi:10.1063/1.3687398.

39. Khan, Mohidus Samad, Deniece Fon, Xu Li, Junfei Tian, John Forsythe, Gil Garnier, and Wei Shen. "Biosurface engineering through ink jet printing." *Colloids and Surfaces B: Biointerfaces* 75, no. 2 (2010): 441–447. doi:10.1016/j.colsurfb.2009.09.032.

40. Li, Xu, Junfei Tian, and Wei Shen. "Progress in patterned paper sizing for fabrication of paper-based microfluidic sensors." *Cellulose* 17, no. 3 (2010): 649–659. doi:10.1007/s10570-010-9401-2.
41. Li, Xu, Junfei Tian, Thanh Nguyen, and Wei Shen. "Paper-based microfluidic devices by plasma treatment." *Analytical chemistry* 80, no. 23 (2008): 9131–9134. doi:10.1021/ac801729t.
42. Olkkonen, Juuso, Kaisa Lehtinen, and Tomi Erho. "Flexographically printed fluidic structures in paper." *Analytical Chemistry* 82, no. 24 (2010): 10246–10250. doi:10.1021/ac1027066.
43. Lu, Yao, Weiwei Shi, Lei Jiang, Jianhua Qin, and Bingcheng Lin. "Rapid prototyping of paper-based microfluidics with wax for low-cost, portable bioassay." *Electrophoresis* 30, no. 9 (2009): 1497–1500. doi:10.1002/elps.200800563.
44. Leung, Vincent, Abdel-Aziz M. Shehata, Carlos D. M. Filipe, and Robert Pelton. "Streaming potential sensing in paper-based microfluidic channels." *Colloids and Surfaces A: Physicochemical and Engineering Aspects* 364, no. 1–3 (2010): 16–18. doi:10.1016/j.colsurfa.2010.04.008.
45. Zhong, Z. W., Z. P. Wang, and G. X. D. Huang. "Investigation of wax and paper materials for the fabrication of paper-based microfluidic devices." *Microsystem Technologies* 18, no. 5 (2012): 649–659. doi:10.1007/s00542-012-1469-í.
46. Chitnis, Girish, Zhenwen Ding, Chun-Li Chang, Cagri A. Savran, and Babak Ziaie. "Laser-treated hydrophobic paper: An inexpensive microfluidic platform." *Lab on a Chip* 11, no. 6 (2011): 1161–1165. doi:10.1039/C0LC00512F.
47. Liu, Hong, Yu Xiang, Yi Lu, and Richard M. Crooks. "Aptamer-based origami paper analytical device for electrochemical detection of adenosine." *Angewandte Chemie* 124, no. 28 (2012): 7031–7034. doi:10.1002/ange.201202929.
48. Lankelma, Jan, Zhihong Nie, Emanuel Carrilho, and George M. Whitesides. "Paper-based analytical device for electrochemical flow-injection analysis of glucose in urine." *Analytical Chemistry* 84, no. 9 (2012): 4147–4152. doi:10.1021/ac3003648.
49. Ratnarathorn, Nalin, Orawon Chailapakul, Charles S. Henry, and Wijitar Dungchai. "Simple silver nanoparticle colorimetric sensing for copper by paper-based devices." *Talanta* 99 (2012): 552–557. doi:10.1016/j.talanta.2012.06.033.
50. Ge, Lei, Jixian Yan, Xianrang Song, Mei Yan, Shenguang Ge, and Jinghua Yu. "Three-dimensional paper-based electrochemiluminescence immunodevice for multiplexed measurement of biomarkers and point-of-care testing." *Biomaterials* 33, no. 4 (2012): 1024–1031. doi:10.1016/j.biomaterials.2011.10.065.
51. Ge, Lei, Shoumei Wang, Xianrang Song, Shenguang Ge, and Jinghua Yu. "3D Origami-based multifunction-integrated immunodevice: Low-cost and multiplexed sandwich chemiluminescence immunoassay on microfluidic paper-based analytical device." *Lab on a Chip* 12, no. 17 (2012): 3150–3158. doi:10.1039/C2LC40325K.
52. Elavarasi, M., A. Rajeshwari, N. Chandrasekaran, and Amitava Mukherjee. "Simple colorimetric detection of Cr^{3+} in aqueous solutions by as synthesized citrate capped gold nanoparticles and development of a paper-based assay." *Analytical Methods* 5, no. 21 (2013): 6211–6218. doi:10.1039/C3AY41435C.
53. Li, Shaoyuan, Xiuhua Chen, Wenhui Ma, Zhao Ding, Cong Zhang, Zhengjie Chen, Xiao He, Yudong Shang, and Yuxin Zou. "An innovative metal ion sensitive "test paper" based on virgin nanoporous silicon wafer: Highly selective to copper (II)." *Scientific Reports* 6, no. 1 (2016): 1–9. doi:10.1038/srep36654.
54. Chen, Guan-Hua, Wei-Yu Chen, Yu-Chun Yen, Chia-Wei Wang, Huan-Tsung Chang, and Chien-Fu Chen. "Detection of mercury (II) ions using colorimetric gold nanoparticles on paper-based analytical devices." *Analytical Chemistry* 86, no. 14 (2014): 6843–6849. doi:10.1021/ac5008688.

55. Pourreza, Nahid, Hamed Golmohammadi, and Saadat Rastegarzadeh. "Highly selective and portable chemosensor for mercury determination in water samples using curcumin nanoparticles in a paper based analytical device." *RSC Advances* 6, no. 73 (2016): 69060–69066. doi:10.1039/C6RA08879A.

56. Nath, Peuli, Ravi Kumar Arun, and Nripen Chanda. "Smart gold nanosensor for easy sensing of lead and copper ions in solution and using paper strips." *RSC Advances* 5, no. 84 (2015): 69024–69031. doi:10.1039/C5RA14886C.

57. Priyadarshni, Nivedita, Peuli Nath, Nagahanumaiah, and Nripen Chanda. "DMSA-functionalized gold nanorod on paper for colorimetric detection and estimation of arsenic (III and V) contamination in groundwater." *ACS Sustainable Chemistry & Engineering* 6, no. 5 (2018): 6264–6272. doi:10.1021/acssuschemeng.8b00068.

58. Song, Shanshan, Shuzhen Zou, Jianping Zhu, Liqiang Liu, and Hua Kuang. "Immunochromatographic paper sensor for ultrasensitive colorimetric detection of cadmium." *Food and Agricultural Immunology* 29, no. 1 (2018): 3–13. doi:10.1080/09540105.2017.1354358.

59. Chen, Guan-Hua, Wei-Yu Chen, Yu-Chun Yen, Chia-Wei Wang, Huan-Tsung Chang, and Chien-Fu Chen. "Detection of mercury(II) ions using colorimetric gold nanoparticles on paper-based analytical devices." *Analytical Chemistry* 86, no. 14 (2014): 6843–6849. doi:10.1021/ac5008688.

60. Zhang, Lina, Yanhu Wang, Chao Ma, Panpan Wang, and Mei Yan. "Self-powered sensor for Hg^{2+} detection based on hollow-channel paper analytical devices." *RSC Advances* 5, no. 31 (2015): 24479–24485. doi:10.1039/C4RA14154G.

61. Feng, Liang, Hui Li, Li-Ya Niu, Ying-Shi Guan, Chun-Feng Duan, Ya-Feng Guan, Chen-Ho Tung, and Qing-Zheng Yang. "A fluorometric paper-based sensor array for the discrimination of heavy-metal ions." *Talanta* 108 (2013): 103–108. doi:10.1016/j.talanta.2013.02.073.

62. Wang, Hu, Ya-jie Li, Jun-feng Wei, Ji-run Xu, Yun-hua Wang, and Guo-xia Zheng. "Paper-based three-dimensional microfluidic device for monitoring of heavy metals with a camera cell phone." *Analytical and Bioanalytical Chemistry* 406, no. 12 (2014): 2799–2807. doi:10.1007/s00216-014-7715-x.

6 Graphene-Based Nanostructures as Plasmonic Nanosensors

6.1 INTRODUCTION

The impact of metal ions on human health and the environment fascinated the measurement of the metal ions using sensors in water [1]. For instance, for maintaining safe water distribution, sensors must make simultaneous measurements at low concentrations of metals in water, such as mercury, lead, and cadmium [2]. When maximum limits of contaminant metal ions are surpassed, the real-time detection competence can confirm a suitable time to take precautions. Graphene-based sensors are comparatively novel to encounter the aim of quick in situ sensing of toxic metal ions in aqueous systems [3–5]. Graphene has fascinated substantial scientific consideration since its initial production through mechanical exfoliation to sense heavy metals [6,7]. The members of the graphene family are reduced graphene oxide (rGO), GO, and pristine graphene. Graphene is like graphite, but it is organized in a systematic hexagonal pattern comprising 2D carbon atoms in a single layer of graphite. A $1\,m^2$ sheet of graphene is relatively light and weighs only $0.77\,mg$ [8]. The IUPAC defines as a single carbon layer of the graphite structure graphene. Thru similarity to a polycyclic aromatic hydrocarbon of quasi-infinite size defines graphene's nature [9]. Graphite already used copiously poses no hazard to the environment, and it is a naturally occurring mineral [10]. The oxygen functional groups, such as carboxyls, hydroxyls, alcohols, and epoxides, are present on the edges and basal planes of oxygenated graphene sheets and GOs that are layered. The oxygen to carbon ratio is around 1:3, confirmed by the chemical analysis [11]. Usually, to large-scale graphene synthesis, GO is perceived as a precursor [12]. The electroreduction [13], chemical reduction [14], enzymatic reduction [15], flash reduction [16], and thermal annealing [17] are the methods by which GO can be reduced to almost graphene. With some structural defects and residual oxygen, GO converts into rGO after reduction.

In contrast to doped conductive polymers, this transformation yields high thermal conductivity higher than Si (36 times), and then As, Ga (100 times) higher [18]. The possible modification of graphene and its classification can be divided into seven routes. By the first route, graphite is oxidized in GO. The stepwise exfoliation of GO is done by the second route. The third route is used to reduce GO into reduced graphene (RGO). The fourth route is produced graphene by mechanical exfoliation of graphite and the oxidation of graphene sheets to GO.

DOI: 10.1201/9781003128281-6

The transformation process is done through routes five and six, which is the thermal decomposition of a SiC wafer, and route seven is the growth of graphene films through chemical vapor deposition (CVD).

6.2 PROPERTIES OF GRAPHENE

Graphene displays several inimitable and outstanding goods. In contrast to CNTs, the zero-band gap of graphene is much higher within the structure and permits its $200,000\,cm^2V^{-1}s^{-1}$ of ultrahigh electron mobility [19]. It has $\approx4,000$ W m^{-1} K^{-1} of outstanding thermal conductivity on lattice symmetry structure-free graphene, owing to the possibility of atomically clean graphene sheets, [20], enormously high surface-to-volume ratio up to $2,600\,m^2g^{-1}$ and [21], ultrahigh capacitance [22], mechanical strength [23,24], and 3,189 S cm^{-1} of outstanding electrical conductivity [25,26]. Besides these exceptional properties, the use of graphene showed the opportunity of the sensitive sensing of numerous toxicants with the exceptionally low electronic noise of graphene [27,28].

6.3 NANOCOMPOSITES FOR COLORIMETRIC RESPONSES

The assemblage of noble NPs such as Pd, Ag, Au with graphene endows them with better catalytic performance and enhances their stability, and these NPs own peroxidase mimetic activity. Several sensors have been proposed to detect heavy metals by employing this feature. For instance, GO-AuNPs nanohybrids can differentiate between single-stranded DNA (ssDNA) and double-stranded DNA (dsDNA). It is based on the metal ion-triggered DNA conformation evolution with peroxidase-like activity. It can detect Pb^{2+} and Hg^{2+} metal ions colorimetrically [29]. Lu et al. demonstrated reduced graphene oxide/polyethylenimine/PdNPs (rGO/PEI/Pd) nanohybrids for the selective and sensitive sensing of Hg^{2+} ions with 0.39 nM of the detection limit. The color of 3,3′,5,5′-tetramethylbenzidine (TMB) significantly enhanced and stimulated the peroxidase mimetic activity in the existence of Hg^{2+} ions through effective oxidation of rGO/PEI/Pd nanohybrids [30]. Correspondingly, to support Au/Fe_3O_4 NPs, Zhi et al. developed MoS_2 aerogels doped with graphene. Here, Fe_3O_4 NPs bestow the magnetic property of aerogels simply for several recurring catalyst recovery and usages. Introducing AuNPs facilitates the detection of Hg^{2+} ions in aqueous systems by making aerogels that possess mercury-stimulated peroxidase mimetic activity. For sensitive colorimetric detection and efficient capture of Hg^{2+}ions, the nanocomposites display suggestively enhanced nanozyme activity due to large adsorption capacity [31]. By adding Hg^{2+} ions, the direct color change of the nanocomposite solution has also been applied for sensing of Hg^{2+} ions besides the peroxidase mimetic activity. Lee's group developed a gold nanocomposite incorporated with multifunctional graphene (G-AuNPs) through a one-pot simple redox reaction. Because of the formation of amalgam AuNPs, G-AuNPs can specifically enrich Hg^{2+} ions in the attendance of ascorbic acid, which induces the change in color of the solution from purple red to light brown, facilitating the "naked-eye" detecting of Hg^{2+} ions [32].

6.4 HEAVY METAL IoN DETECTION

Through food chains, heavy metal ions accumulated in the human body, such as lead (Pb^{2+}), arsenic (As^{3+}), mercury (Hg^{2+}), chromium (Cr^{6+} or Cr^{3+}), and cadmium (Cd^{2+}) ions. These non-biodegradable and highly lethal contaminants might have opposing effects on the reproductive, central nervous, and immune systems [33]. Food and water are vital to manufacturing novel analytical methods in the environment, food, and water [34]. In contrast to other optical and spectroscopic methods, the electrochemical method facilitates real-time monitoring of pollutants, a novel, economical technique [35]. For the monitoring of environmental contaminants efficiently, diverse sorts of nanostructured materials have been developed. From aqueous solutions to remove organic dyes and inorganic pollutants, a vast potential has been shown by graphene-based nanocomposites [36,37]. Table 6.1 summarizes the electrochemical sensors based on graphene for the sensing of various metal ions. For the rapid recognition of Cu^{2+}, Hg^{2+}, Pb^{2+}, and Cd^{2+} ions, L-cysteine-functionalized GO was used by Muralikrishna et al. [38]. The developed detection probe showed the 0.4–1.2 mM of linear detection range for Pb^{2+} ions, and Hg^{2+}, Cu^{2+}, and Cd^{2+} were 0.4–2.0 mM. The L-cysteine/GO composite-based sensor for sensing Cu^{2+}, Pb^{2+}, Hg^{2+}, and Cd^{2+} ions exhibited LODs of 0.261, 0.416, 1.113, and 0.366 mg L^{-1} found to be lower than the World Health Organization (WHO) standards.

A GO nanocomposite and ruthenium(II) bipyridine complex ($[Ru(bpy)_3]^{2+}$) have been utilized to sense Pb^{2+}, Cd^{2+}, Hg^{2+} and As^{3+} metals ions by Gumpu et al. [46]. The LOD of the developed sensor was 1.41, 2.8, 1.6, and 2.3 nM for Pb^{2+}, Cd^{2+}, Hg^{2+}, and As^{3+} metal ions.

In a similar study, for rapid detection of Cd^{2+} and Pb^{2+} ions, the use of TiO_2–graphene hybrid nanostructures was reported by Zhang et al. [56]. In the linear ranges from 6.0×10^{-7} to 3.2×10^{-5} M and 1.0×10^{-8} to 3.2×10^{-5} M, the developed sensor exhibited LOD at 2.0×10^{-9} and 1.0×10^{-10} M for Cd^{2+} and Pb^{2+} ions, correspondingly.

The electrochemical determination of Hg^{2+} ions have been done using modified screen-printed carbon electrodes based on activated graphite [57]. The sensor showed a sensitivity of 81.5 mA $ppm^{-1}cm^{-2}$ with a LOD of 4.6 ppb and a linear range of 0.05–14.77 ppm. For the detection of Hg^{2+} ions in aqueous samples, the LOD value is significantly lower than the permissible limit.

For the electrochemical sensing of Pb^{2+} ions in aqueous samples, carbonyldiimidazole was used as a cross-linker to change cysteine-functionalized GO by Seenivasan et al. [58]. In 1.4–28 ppb of linear range, the developed sensor showed 0.07 ppb of LOD value toward Pb^{2+} ions, having S/N=3 ratios. The permissible limit of Pb^{2+} ions for drinking water set by WHO this value is 2-fold lower in contrast to the threshold of 10 ppb.

TABLE 6.1

The Detection of Inorganic Pollutants on Diverse Graphene-Based Colorimetric/Electrochemical Sensors

Graphene-Based Nanocomposites	Detection Method	Metal Ions	LOD	Detection Range	References
AuNPs/GR/cysteine composite	SWASV	Cd^{2+} Pb^{2+}	$0.1\,\mu g\,L^{-1}$ $0.05\,\mu g\,L^{-1}$	$0.5–40\,mg\,L^{-1}$	[39]
GO/AgNPs composite	SWV	Pb^{2+}	80 pM	0.1 nM to 10 µM	[40]
Pi-A/rGO nanocomposite	SWASV	Cu^{2+}	$0.67\,\mu g\,L^{-1}$	$5–300\,\mu g\,L^{-1}$	[41]
GR-Au modified electrode	CV, SWV	Hg^{2+}	0.001 aM	1 aM to 100 nM	[42]
L-Cysteine/AuNPs/rGO	DPSV	Cu^{2+}	$0.037\,\mu g\,mL^{-1}$	$2–60\,\mu g\,mL^{-1}$	[43]
GQDs@VMSF	DPV	Hg^{2+} Cu^{2+} Cd^{2+}	9.8 pM 8.3 pM 4.3 nM	10 pM to 1 nM 10 pM to 1 nM 20 nM to 1 µM	[44]
DNA-modified 3D rGO/CS composite	EIS	Hg^{2+}	0.016 mM	0.1 – 10 nM	[45]
Ruthenium(II) bipyridine $[Ru(bpy)_3]^{2+}$ complex/GO composite	CV, EIS	Cd^{2+} Pb^{2+} As^{2+} Hg^{2+}	2.8 nM 1.41 nM 2.3 nM 1.6 nM	0.5–0.3 µM 0.05–0.25 µM 0.05–1.8 µM 0.1–1.2 µM	[46]
GO modified Au electrode	SWV	Pb^{2+} Hg^{2+} Cu^{2+}	0.4 ppb 0.8 ppb 1.2 ppb	0–50 ppb 0–15 ppb 0–200 ppb	[47]
Heparin modified CS/GR composite	SWASV	Pb^{2+}	$0.03\,\mu g\,L^{-1}$	$1.125–8.25\,\mu g\,L^{-1}$	[48]
$MnFe_2O_4$/GO nanocomposite	SWASV	Pb^{2+}	0.0883 µM	0.2–1.1 µM	[49]
GR nanodots/porous Au electrode	SWV	Pb^{2+} Cu^{2+}	0.8 nM 1 nM	0.006–2.5 µM 0.009–4 µM	[50]
Hollow AuPd-flower-like MnO_2–hemin@rGO composite	DPV	Pb^{2+}	0.034 pM	0.1 pM to 200 nM	[51]
rGO-CS/poly-L-lysine nanocomposite	DPASV	Cd^{2+} Pb^{2+} Cu^{2+}	$0.01\,\mu g\,mL^{-1}$ $0.02\,\mu g\,mL^{-1}$ $0.02\,\mu g\,mL^{-1}$	$0.05–10\,\mu g\,mL^{-1}$	[52]
rGO/AuNPs composite	DPV	Fe^{3+}	3.5 nM	30–3,000 nM	[53]
AuNPs/rGO nanocomposite	CV, EIS	Hg^{2+}	$1.5\,ng\,L^{-1}$	$10\,ng\,L^{-1}$ to $1.0\,\mu g\,L^{-1}$	[54]
Nile red-GR composite	CV, DPV	Fe^{3+}	24.9 µM	30–1,000 µM	[55]

For Hg^{2+} ion detection in the aqueous sample, Zhang et al. [42] developed a graphene electrode modified by electrodeposited Au by activating thymine–Hg^{2+}–thymine (T–Hg^{2+}–T) coordination chemistry for amplification of signal with a LOD of 0.001 mM. Among numerous mixtures of interfering metal ions with Hg^{2+} ions, the developed sensor selectively detects Hg^{2+} ions. The square wave voltammograms (SWV) have been displayed as a function of Hg^{2+} ions concentration of the biosensor. In this sensor, design to additional advancement of the signal response, a DNA probe rich in 10-mer thymine (P_1) functionalized on a glassy carbon electrode (GCE) surface with electrodeposited GR-Au and the DNA probe rich in 29-mer guanine (P_3) functionalized Au nanocarrier labeled with an MB. For P_2 hybridization, owing to the T–Hg^{2+}–T coordination chemistry and MB-nano-Au-P 3s, the T–T mismatched dsDNA forms of the electrode surface with the addition of Hg^{2+} among a DNA probe rich in 22-mer thymine (P_2) and P_1. In the concentration range from 10 mm to 0.001 nM, the SWV of the Hg^{2+} sensor exhibited LOD of 0.001 nM toward sensing of Hg^{2+} ions. The SWV measurements showed that the developed biosensor toward Hg^{2+} ions also showed inflated selectivity in the existence of Ca^{2+}, K^+, Cd^{2+}, Ba^{2+}, Co^{2+}, Cr^{2+}, Mg^{2+}, Cu^{2+}, Pb^{2+}, Mn^{2+}, Zn^{2+}, Ni^{2+}, Al^{3+}, and Fe^{3+} metal ions of 500 nM. The developed sensor showed an LOD of 10 nM for Hg^{2+} ions with a sensitivity value <5% in the presence of other metals.

Yu et al. [59] developed an rGO sensor modified noncovalently to detect Hg^{2+} to the picomolar. The LOD of the developed sensor was less than set by USEPA and WHO, which was found to be 0.1 nM for Hg^{2+} ions.

For the sensing of Cd^{2+}, Cu^{2+}, and Hg^{2+} ions, Lu et al. [44], using silica nanochannel confined graphene quantum dots (GQDs) and dopamine, developed an electrochemical sensor. For the sensing of Cd^{2+}, Cu^{2+}, and Hg^{2+} ions, the NH_2-GQD@VMSF/ITO electrodes, OH-GQD@VMSF/ITO, and vertically ordered mesoporous silica nanochannel film (VMSF)/ITO electrode were used. In contrast to VMSF/ITO electrode having no GQDs used to detect Cd^{2+}, Cu^{2+}, and Hg^{2+} ions, both NH_2-GQD@VMSF/ITO and OHGQD@VMSF/ITO electrodes showed substantially augmented current signals. Because of their molecular interactions with Cd^{2+}, Cu^{2+}, and Hg^{2+} ions, high specific surface area inherited thru GQDs besides the effective transfer of electrons among metal and GQDs, a substantial upsurge in current signals is allied with the GQDs. The DPV curves of NH_2-GQD@VMSF/ITO and OH-GQD@VMSF/ITO electrodes have been recorded as a function of diverse concentrations of Cd^{2+}, Cu^{2+}, and Hg^{2+}, respectively. Using VMSF/ITO electrode, the LOD for Cu^{2+} and Hg^{2+} ions was found to be 32 nM and 0.3 mM, correspondingly in contrast to the OH-GQD@VMSF/ITO electrode-based sensing platform, which can detect Cu^{2+} and Hg^{2+} ions at concentrations as low as 8.3 and 9.8 pM, respectively. Similarly, for detecting the Cd^{2+} ions, the NH_2-GQD@VMSF/ITO electrode showed LOD of 4.3 nM; apart from this, it is found up to 120 nM in the case of the VMSF/ITO electrode. Using the electrode made by incorporating OH-GQDs, the LOD was reduced through five orders of magnitude for detecting Hg^{2+} ions. To detect dopamine, the ensuing OH-GQD@VMSF/Au electrode exhibited LOD of 120 nm after replacing the ITO electrode with Au electrode and showed a linear range of 200 nm to 100 mM. In the presence of other interfering metal ions, the GQD@VMSF/ITO electrode-based sensors exhibited high selectivity and sensitivity toward Cd^{2+}, Cu^{2+}, and Hg^{2+} ions, for example, in Mg^{2+}, Co^{2+}, Fe^{3+}, Ca^{2+}, BSA, hemoglobin, and Cr^{3+}. Likewise,

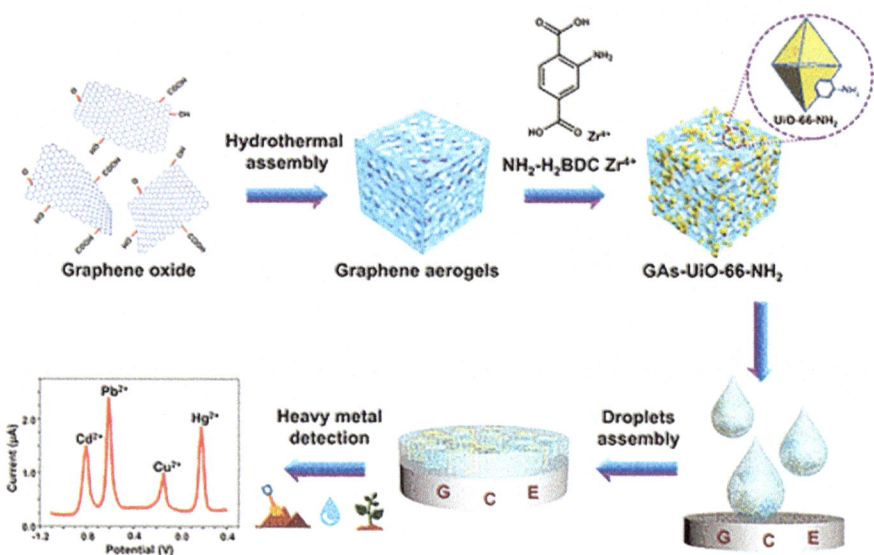

FIGURE 6.1   The sensing mechanism of toxic metals on graphene-based nanostructures. (Reprinted with permission from Ref. [12]. Copyright 2018 American Chemical Society.)

toward dopamine, high selectivity is exhibited by the OH-GQD@VMSF/Au electrode-based sensor in the existence of other co-bio analytes such as amino acids, common ions, ascorbic acid, uric acid, and protein (concanavalin A). Figure 6.1 shows the schematic sensing mechanism of toxic metals using graphene-based nanostructures.

6.5   FLUORESCENCE PROBES USING GQDs

The rational elucidation is that the FWHM is wide with asymmetric emission peak, large graphene or GO lack a bandgap, and the fluorescence QY is usually low. It is dependent on its NP size and surface status, while, as demonstrated above, optical fluorescence has been provisionally detected in GO. Considerable effort has been dedicated toward the "top-down" method, which is to develop 0D GQDs from the GO sheets or 2D graphene [60–62], or the "bottom-up" strategy is where GQDs were developed thru pyrolysis of structure-defined small molecules [63–66]. Significantly, in the determination of heavy metal ions, doping GQDs with heteroatoms (e.g., S, N, and B) may efficiently tune their intrinsic properties, such as local chemical features, surface, and optical characteristics simplifying their benefits in chemical sensing as fluorescent probes.

Shen et al. developed graphene quantum dots (S, N-GQDs) co-doped with sulfur and nitrogen in a hydrothermal process based on one-step bottom-up molecular fusion. Although in a mixture of metal ions, the prepared S, N-GQDs could similarly detect the Ag^+, Cu^{2+}, and Fe^{3+} ions with comparatively low sensing limits. The detection mechanism was done through the specific and differential sensing of the Ag^+, Cu^{2+}, and Fe^{3+} ions depending on the masking agents on the fluorescence of S, N-GQDs, and the quenching effect of metals [67].

Using the dual-excitation GQDs, Xu and co-workers reported the detection of Hg^{2+} and Fe^{3+} ions by synthesizing a masking agent-free technique [68]. Fascinatingly, one of the fluorescence channels could be quenched by Hg^{2+} ions while the other one responded to Fe^{3+} ions in a single matrix. Under two different excitation wavelengths, the prepared GQDs could emit strong blue fluorescence for concurrent sensing of these two metal ions. In specific sensing of heavy metal ions using doped GQDs or bare GQDs, there are many other reports on the "top-down" or "bottom-up" synthesis approach [69,70]. The "turn off" mechanism grieves the false-positive results and limitation of poor selectivity. Most of the approaches rely on using the direct interaction among metal ions and GQDs. To determine the Hg^{2+} ions, the drawback of single signal "turn off" responsive methods overwhelmed through the synthesis of a ratiometric probe. For the built-in correction of ecological belongings, the silica nanosphere incorporating the red-emitting CdTe QDs act as the internal standard in this detection method.

To detect the Hg^{2+} ions through electrostatic interaction, the GQDs serve as reaction sites self-assembled on the silica surface. The detection of Hg^{2+} ions was done thru the solution fluorescence color, which unceasingly altered from blue to red because of the fluorescence of GQDs. In contrast, the fluorescence of CdTe QDs was preserved [71].

In contrast, as a dual-fluorescence probe for detecting Cu^{2+} ions, GSH-functionalized GQDs and CdTe QDs have been developed, acting as an internal standard and as the responsive probe in the detection process, correspondingly [72]. To detect Cu^{2+} ions, the blue fluorescence of GQDs remained constant while the red fluorescence of CdTe QDs can be quenched upon the addition of Cu^{2+} in the ratiometric response.

In a similar study, Zhao and co-workers developed a ratiometric fluorescence approach for sensing Ag^+ ions. Therefore, they used o-phenylenediamine (OPD) and retained GQDs as the specific recognition probe and reference fluorophore, respectively [73]. The fluorescence of GQDs at 445 nm would be concurrently quenched through the generated 2,3-diaminophenazine to produce 2,3-diaminophenazine through fluorescence resonance energy transfer (FRET). In comparison, with a strong fluorescence emission at 557 nm OPD could be oxidized upon the addition of Ag^+ ions.

6.6 CONCLUDING REMARKS AND FUTURE PERSPECTIVES

In many fields of benefits, especially confirming health and environmental protection, inorganic pollutants' selective and sensitive sensing is in vast consideration. The graphene-based nanomaterials have improved the recital of sensors for detecting metal ions because of their optical properties. Such as high surface area to load other species, bright fluorescence, efficient fluorescence quenching, and photocatalysis. This chapter has elaborated graphene-based nanomaterials for the optical sensing of various inorganic pollutants such as many heavy metals.

In this, to comprise the diverse categories of graphene-based nanomaterials (comprising GQDs, GO, and graphene), comprehensive information has been provided for heavy metal ion detection. For the location of NP and molecule binding, GO, or graphene is a perfect substrate.

Similarly, for developing a metal ion sensing platform, GO, or graphene-based nanocomposites, is an effective fluorescence quencher that has been extensively useful. So far, because of limited detection sensitivity and low fluorescence quantum yield, some efforts are dedicated to using GO or graphene as fluorescence or colorimetric detection probes. To advance the selectivity of the 0D GQDs, which work as fluorescence probes for modification with special ligands (comprising large biomolecules and small molecules) or direct "turn off" detection of inorganic pollutants. Under a range of strict ecological conditions, the synthesis of sensors could offer high specificity, which should be placed to detect heavy metals. The researchers are designing concurrent and multiple metal ion sensing methods. The future developments are intended as additional vital elements, especially in colorimetric metal ion or optically fluorescent nanosensors.

REFERENCES

1. Fairbrother, Anne, Randall Wenstel, Keith Sappington, and William Wood. "Framework for metals risk assessment." *Ecotoxicology and Environmental Safety* 68, no. 2 (2007): 145–227. doi:10.1016/j.ecoenv.2007.03.015.
2. Aragay, Gemma, Josefina Pons, and Arben Merkoçi. "Recent trends in macro-, micro-, and nanomaterial-based tools and strategies for heavy-metal detection." *Chemical Reviews* 111, no. 5 (2011): 3433–3458. doi:10.1021/cr100383r.
3. Singh, Virendra, Daeha Joung, Lei Zhai, Soumen Das, Saiful I. Khondaker, and Sudipta Seal. "Graphene based materials: Past, present and future." *Progress in Materials Science* 56, no. 8 (2011): 1178–1271. doi:10.1016/j.pmatsci.2011.03.003.
4. Raza, Hassan, and Edwin C. Kan. "Armchair graphene nanoribbons: Electronic structure and electric-field modulation." *Physical Review B* 77, no. 24 (2008): 245434. doi:10.1103/PhysRevB.77.245434.
5. Jing, Zhao, Zhang Guang-Yu, and Shi Dong-Xia. "Review of graphene-based strain sensors." *Chinese Physics B* 22, no. 5 (2013): 057701. doi:10.1088/1674-1056/22/5/057701.
6. Novoselov, Kostya S., Andre K. Geim, Sergei V. Morozov, Dingde Jiang, Yanshui Zhang, Sergey V. Dubonos, Irina V. Grigorieva, and Alexandr A. Firsov. "Electric field effect in atomically thin carbon films." *Science* 306, no. 5696 (2004): 666–669. doi:10.1126/science.1102896.
7. Dresselhaus, Mildred S., and Paulo T. Araujo. "Perspectives on the 2010 nobel prize in physics for graphene." (2010): 6297–6302. doi:10.1021/nn1029789.
8. Geim, Andre Konstantin. "Graphene: Status and prospects." *Science* 324, no. 5934 (2009): 1530–1534. doi:10.1126/science.1158877.
9. Fitzer, E., K-H. Kochling, H. P. Boehm, and H. Marsh. "Recommended terminology for the description of carbon as a solid (IUPAC Recommendations 1995)." *Pure and Applied Chemistry* 67, no. 3 (1995): 473–506.
10. Salas, Everett C., Zhengzong Sun, Andreas Lüttge, and James M. Tour. "Reduction of graphene oxide via bacterial respiration." *ACS Nano* 4, no. 8 (2010): 4852–4856. doi:10.1021/nn101081t.

11. Stankovich, Sasha, Richard D. Piner, Xinqi Chen, Nianqiang Wu, SonBinh T. Nguyen, and Rodney S. Ruoff. "Stable aqueous dispersions of graphitic nanoplatelets via the reduction of exfoliated graphite oxide in the presence of poly (sodium 4-styrenesulfonate)." *Journal of Materials Chemistry* 16, no. 2 (2006): 155–158. doi:10.1039/B512799H.

12. Lu, Muxin, Yajuan Deng, Yi Luo, Junping Lv, Tianbao Li, Juan Xu, Shu-Wei Chen, and Jinyi Wang. "Graphene aerogel–metal–organic framework-based electrochemical method for simultaneous detection of multiple heavy-metal ions." *Analytical Chemistry* 91, no. 1 (2018): 888–895. doi:10.1021/acs.analchem.8b03764.

13. Hilder, Matthias, Bjorn Winther-Jensen, Dan Li, Maria Forsyth, and Douglas R. MacFarlane. "Direct electro-deposition of graphene from aqueous suspensions." *Physical Chemistry Chemical Physics* 13, no. 20 (2011): 9187–9193. doi:10.1039/C1CP20173E.

14. Park, Sungjin, and Rodney S. Ruoff. "Chemical methods for the production of graphenes." *Nature Nanotechnology* 4, no. 4 (2009): 217–224. doi:10.1038/nnano.2009.58.

15. Quinlan, Ronald A., Artjay Javier, Edward E. Foos, Leonard Buckley, Mingyao Zhu, Kun Hou, Erika Widenkvist, Martin Drees, Ulf Jansson, and Brian C. Holloway. "Transfer of carbon nanosheet films to nongrowth, zero thermal budget substrates." *Journal of Vacuum Science & Technology B, Nanotechnology and Microelectronics: Materials, Processing, Measurement, and Phenomena* 29, no. 3 (2011): 030602. doi:10.1116/1.3574524.

16. Cote, Laura J., Rodolfo Cruz-Silva, and Jiaxing Huang. "Flash reduction and patterning of graphite oxide and its polymer composite." *Journal of the American Chemical Society* 131, no. 31 (2009): 11027–11032. doi:10.1021/ja902348k.

17. McAllister, Michael J., Je-Luen Li, Douglas H. Adamson, Hannes C. Schniepp, Ahmed A. Abdala, Jun Liu, Margarita Herrera-Alonso et al. "Single sheet functionalized graphene by oxidation and thermal expansion of graphite." *Chemistry of Materials* 19, no. 18 (2007): 4396–4404. doi:10.1021/cm0630800.

18. Bao, Qiaoliang, and Kian Ping Loh. "Graphene photonics, plasmonics, and broadband optoelectronic devices." *ACS Nano* 6, no. 5 (2012): 3677–3694. doi:10.1021/nn300989g.

19. Hong, Seunghun, and Sung Myung. "A flexible approach to mobility." *Nature Nanotechnology* 2, no. 4 (2007): 207–208. doi:10.1038/nnano.2007.89.

20. Chen, Shanshan, Qingzhi Wu, Columbia Mishra, Junyong Kang, Hengji Zhang, Kyeongjae Cho, Weiwei Cai, Alexander A. Balandin, and Rodney S. Ruoff. "Thermal conductivity of isotopically modified graphene." *Nature Materials* 11, no. 3 (2012): 203–207. doi:10.1038/nmat3207.

21. Ishigami, Masa, J. H. Chen, W. G. Cullen, M. S. Fuhrer, and E. D. Williams. "Atomic structure of graphene on SiO$_2$." *Nano Letters* 7, no. 6 (2007): 1643–1648. doi:10.1021/nl070613a.

22. Sridhar, Vadahanambi, Hyun-Jun Kim, Jung-Hwan Jung, Changgu Lee, Sungjin Park, and Il-Kwon Oh. "Defect-engineered three-dimensional graphene–nanotube–palladium nanostructures with ultrahigh capacitance." *ACS Nano* 6, no. 12 (2012): 10562–10570. doi:10.1021/nn3046133.

23. Chen, Jian, Hui Bi, Shengrui Sun, Yufeng Tang, Wei Zhao, Tianquan Lin, Dongyun Wan et al. "Highly conductive and flexible paper of 1D silver-nanowire-doped graphene." *ACS Applied Materials & Interfaces* 5, no. 4 (2013): 1408–1413. doi:10.1021/am302825w.

24. Layek, Rama K., Atanu Kuila, Dhruba P. Chatterjee, and Arun K. Nandi. "Amphiphilic poly (N-vinyl pyrrolidone) grafted graphene by reversible addition and fragmentation polymerization and the reinforcement of poly (vinyl acetate) films." *Journal of Materials Chemistry A* 1, no. 36 (2013): 10863–10874. doi:10.1039/C3TA11853C.

25. Bolotin, Kirill I., K. J. Sikes, Zd Jiang, M. Klima, G. Fudenberg, J. ea Hone, Ph Kim, and H. L. Stormer. "Ultrahigh electron mobility in suspended graphene." *Solid State Communications* 146, no. 9–10 (2008): 351–355. doi:10.1016/j.ssc.2008.02.024.

26. Hwang, JongSeung, Duk Soo Kim, Doyeol Ahn, and Sung Woo Hwang. "Transport properties of a DNA-conjugated single-wall carbon nanotube field-effect transistor." *Japanese Journal of Applied Physics* 48, no. 6S (2009): 06FD08. doi:10.1143/JJAP.48.06FD08.

27. Joshi, Rakesh K., Humberto Gomez, Farah Alvi, and Ashok Kumar. "Graphene films and ribbons for sensing of O_2, and 100 ppm of CO and NO_2 in practical conditions." *The Journal of Physical Chemistry C* 114, no. 14 (2010): 6610–6613. doi:10.1021/jp100343d.

28. Kim, Keun Soo, Yue Zhao, Houk Jang, Sang Yoon Lee, Jong Min Kim, Kwang S. Kim, Jong-Hyun Ahn, Philip Kim, Jae-Young Choi, and Byung Hee Hong. "Large-scale pattern growth of graphene films for stretchable transparent electrodes." *Nature* 457, no. 7230 (2009): 706–710. doi:10.1038/nature07719.

29. Chen, Xia, Niu Zhai, John Hugh Snyder, Qiansi Chen, Pingping Liu, Lifeng Jin, Qingxia Zheng, Fucheng Lin, Jiming Hu, and Huina Zhou. "Colorimetric detection of Hg^{2+} and Pb^{2+} based on peroxidase-like activity of graphene oxide–gold nanohybrids." *Analytical Methods* 7, no. 5 (2015): 1951–1957. doi:10.1039/C4AY02801E.

30. Zhang, Shouting, Dongxu Zhang, Xuehong Zhang, Denghui Shang, Zhonghua Xue, Duoliang Shan, and Xiaoquan Lu. "Ultratrace naked-eye colorimetric detection of Hg^{2+} in wastewater and serum utilizing mercury-stimulated peroxidase mimetic activity of reduced graphene oxide-PEI-Pd nanohybrids." *Analytical Chemistry* 89, no. 6 (2017): 3538–3544. doi:10.1021/acs.analchem.6b04805.

31. Zhi, Lihua, Wei Zuo, Fengjuan Chen, and Baodui Wang. "3D MoS_2 composition aerogels as chemosensors and adsorbents for colorimetric detection and high-capacity adsorption of Hg^{2+}." *ACS Sustainable Chemistry & Engineering* 4, no. 6 (2016): 3398–3408. doi:10.1021/acssuschemeng.6b00409.

32. Yan, Zhengquan, Hongtao Xue, Karsten Berning, Yun-Wah Lam, and Chun-Sing Lee. "Identification of multifunctional graphene–gold nanocomposite for environment-friendly enriching, separating, and detecting Hg^{2+} simultaneously." *ACS Applied Materials & Interfaces* 6, no. 24 (2014): 22761–22768. doi:10.1021/am506875t.

33. Aragay, Gemma, Josefina Pons, and Arben Merkoçi. "Recent trends in macro-, micro-, and nanomaterial-based tools and strategies for heavy-metal detection." *Chemical Reviews* 111, no. 5 (2011): 3433–3458. doi:10.1021/cr100383r.

34. Fu, Fenglian, and Qi Wang. "Removal of heavy metal ions from wastewaters: A review." *Journal of Environmental Management* 92, no. 3 (2011): 407–418. doi:10.1016/j.jenvman.2010.11.011.

35. Bansod, BabanKumar, Tejinder Kumar, Ritula Thakur, Shakshi Rana, and Inderbir Singh. "A review on various electrochemical techniques for heavy metal ions detection with different sensing platforms." *Biosensors and Bioelectronics* 94 (2017): 443–455. doi:10.1016/j.bios.2017.03.031.

36. Gong, Xuezhong, Guozhen Liu, Yingshun Li, Denis Yau Wai Yu, and Wey Yang Teoh. "Functionalized-graphene composites: Fabrication and applications in sustainable energy and environment." *Chemistry of Materials* 28, no. 22 (2016): 8082–8118. doi:10.1021/acs.chemmater.6b01447.

37. Vilela, Diana, Jemish Parmar, Yongfei Zeng, Yanli Zhao, and Samuel Sánchez. "Graphene-based microbots for toxic heavy metal removal and recovery from water." *Nano Letters* 16, no. 4 (2016): 2860–2866. doi:10.1021/acs.nanolett.6b00768.

38. Muralikrishna, S., K. Sureshkumar, Thomas S. Varley, Doddahalli H. Nagaraju, and Thippeswamy Ramakrishnappa. "In situ reduction and functionalization of graphene oxide with L-cysteine for simultaneous electrochemical determination of cadmium (II), lead (II), copper (II), and mercury (II) ions." *Analytical Methods* 6, no. 21 (2014): 8698–8705. doi:10.1039/C4AY01945H.

39. Zhu, Lian, Lili Xu, Baozhen Huang, Ningming Jia, Liang Tan, and Shouzhuo Yao. "Simultaneous determination of Cd (II) and Pb (II) using square wave anodic stripping voltammetry at a gold nanoparticle-graphene-cysteine composite modified bismuth film electrode." *Electrochimica Acta* 115 (2014): 471–477. doi:10.1016/j.electacta.2013.10.209.

40. Tang, Shurong, Ping Tong, Xiuhua You, Wei Lu, Jinghua Chen, Guangwen Li, and Lan Zhang. "Label free electrochemical sensor for Pb^{2+} based on graphene oxide mediated deposition of silver nanoparticles." *Electrochimica Acta* 187 (2016): 286–292. doi:10.1016/j.electacta.2015.11.040.

41. Yang, Liuqing, Na Huang, Liyan Huang, Meiling Liu, Haitao Li, Youyu Zhang, and Shouzhuo Yao. "An electrochemical sensor for highly sensitive detection of copper ions based on a new molecular probe Pi-A decorated on graphene." *Analytical Methods* 9, no. 4 (2017): 618–624. doi:10.1039/C6AY03006H.

42. Zhang, Yi, Guang Ming Zeng, Lin Tang, Jun Chen, Yuan Zhu, Xiao Xiao He, and Yan He. "Electrochemical sensor based on electrodeposited graphene-Au modified electrode and nano Au carrier amplified signal strategy for attomolar mercury detection." *Analytical Chemistry* 87, no. 2 (2015): 989–996. doi:10.1021/ac503472p.

43. Yiwei, Xu, Zhang Wen, Huang Xiaowei, Shi Jiyong, Zou Xiaobo, Li Yanxiao, Cui Xueping, Haroon Elrasheid Tahir, and Li Zhihua. "A self-assembled L-cysteine and electrodeposited gold nanoparticles-reduced graphene oxide modified electrode for adsorptive stripping determination of copper." *Electroanalysis* 30, no. 1 (2018): 194–203. doi:10.1002/elan.201700637.

44. Lu, Lili, Lin Zhou, Jie Chen, Fei Yan, Jiyang Liu, Xiaoping Dong, Fengna Xi, and Peng Chen. "Nanochannel-confined graphene quantum dots for ultrasensitive electrochemical analysis of complex samples." *ACS Nano* 12, no. 12 (2018): 12673–12681. doi:10.1021/acsnano.8b07564.

45. Zhang, Zhihong, Xiaoming Fu, Kunzhen Li, Ruixue Liu, Donglai Peng, Linghao He, Minghua Wang, Hongzhong Zhang, and Liming Zhou. "One-step fabrication of electrochemical biosensor based on DNA-modified three-dimensional reduced graphene oxide and chitosan nanocomposite for highly sensitive detection of Hg (II)." *Sensors and Actuators B: Chemical* 225 (2016): 453–462. doi:10.1016/j.snb.2015.11.091.

46. Gumpu, Manju Bhargavi, Murugan Veerapandian, Uma Maheswari Krishnan, and John Bosco Balaguru Rayappan. "Simultaneous electrochemical detection of Cd (II), Pb (II), As (III) and Hg (II) ions using ruthenium (II)-textured graphene oxide nanocomposite." *Talanta* 162 (2017): 574–582. doi:10.1016/j.talanta.2016.10.076.

47. YangáTeoh, Wey. "Graphene oxide-based electrochemical sensor: A platform for ultrasensitive detection of heavy metal ions." *RSC Advances* 4, no. 47 (2014): 24653–24657. doi:10.1039/C4RA02247E.

48. Priya, T., N. Dhanalakshmi, and N. Thinakaran. "Electrochemical behavior of Pb (II) on a heparin modified chitosan/graphene nanocomposite film coated glassy carbon electrode and its sensitive detection." *International Journal of Biological Macromolecules* 104 (2017): 672–680. doi:10.1016/j.ijbiomac.2017.06.082.

49. Zhou, Shao-Feng, Xiao-Juan Han, Hong-Lei Fan, Jin Huang, and Ya-Qing Liu. "Enhanced electrochemical performance for sensing Pb (II) based on graphene oxide incorporated mesoporous MnFe$_2$O$_4$ nanocomposites." *Journal of Alloys and Compounds* 747 (2018): 447–454. doi:10.1016/j.jallcom.2018.03.037.

50. Zhu, Huihui, Yuanhong Xu, Ao Liu, Na Kong, Fukai Shan, Wenrong Yang, Colin J. Barrow, and Jingquan Liu. "Graphene nanodots-encaged porous gold electrode fabricated via ion beam sputtering deposition for electrochemical analysis of heavy metal ions." *Sensors and Actuators B: Chemical* 206 (2015): 592–600. doi:10.1016/j.snb.2014.10.009.

51. Xue, Shuyan, Pei Jing, and Wenju Xu. "Hemin on graphene nanosheets functionalized with flower-like MnO₂ and hollow Au Pd for the electrochemical sensing lead ion based on the specific DNAzyme." *Biosensors and Bioelectronics* 86 (2016): 958–965. doi:10.1016/j.bios.2016.07.111.

52. Guo, Zhuo, Xian-ke Luo, Ya-hui Li, Qi-Nai Zhao, Meng-meng Li, Yang-ting Zhao, Tian-shuai Sun, and Chi Ma. "Simultaneous determination of trace Cd (II), Pb (II) and Cu (II) by differential pulse anodic stripping voltammetry using a reduced graphene oxide-chitosan/poly-l-lysine nanocomposite modified glassy carbon electrode." *Journal of Colloid and Interface Science* 490 (2017): 11–22. doi:10.1016/j.jcis.2016.11.006.

53. Zhu, Yun, Dawei Pan, Xueping Hu, Haitao Han, Mingyue Lin, and Chenchen Wang. "An electrochemical sensor based on reduced graphene oxide/gold nanoparticles modified electrode for determination of iron in coastal waters." *Sensors and Actuators B: Chemical* 243 (2017): 1–7. doi:10.1016/j.snb.2016.11.108.

54. Wang, Nan, Meng Lin, Hongxiu Dai, and Houyi Ma. "Functionalized gold nanoparticles/reduced graphene oxide nanocomposites for ultrasensitive electrochemical sensing of mercury ions based on thymine–mercury–thymine structure." *Biosensors and Bioelectronics* 79 (2016): 320–326. doi:10.1016/j.bios.2015.12.056.

55. Sadak, Omer, Ashok K. Sundramoorthy, and Sundaram Gunasekaran. "Highly selective colorimetric and electrochemical sensing of iron (III) using Nile red functionalized graphene film." *Biosensors and Bioelectronics* 89 (2017): 430–436. doi:10.1016/j.bios.2016.04.073.

56. Zhang, Hongfen, Shaomin Shuang, Guizhen Wang, Yujing Guo, Xili Tong, Peng Yang, Anjia Chen, Chuan Dong, and Yong Qin. "TiO₂–graphene hybrid nanostructures by atomic layer deposition with enhanced electrochemical performance for Pb (II) and Cd (II) detection." *RSC advances* 5, no. 6 (2015): 4343–4349. doi:10.1039/C4RA09779C.

57. KannanáRamaraj, Sayee. "A highly sensitive and selective electrochemical determination of Hg (II) based on an electrochemically activated graphite modified screen-printed carbon electrode." *Analytical Methods* 6, no. 20 (2014): 8368–8373. doi:10.1039/C4AY01805B.

58. Seenivasan, Rajesh, Woo-Jin Chang, and Sundaram Gunasekaran. "Highly sensitive detection and removal of lead ions in water using cysteine-functionalized graphene oxide/polypyrrole nanocomposite film electrode." *ACS Applied Materials & Interfaces* 7, no. 29 (2015): 15935–15943. doi:10.1021/acsami.5b03904.

59. Yu, Chunmeng, Yunlong Guo, Hongtao Liu, Ni Yan, Zhiyan Xu, Gui Yu, Yu Fang, and Yunqi Liu. "Ultrasensitive and selective sensing of heavy metal ions with modified graphene." *Chemical Communications* 49, no. 58 (2013): 6492–6494. doi:10.1039/C3CC42377H.

60. Pan, Dengyu, Jingchun Zhang, Zhen Li, and Minghong Wu. "Hydrothermal route for cutting graphene sheets into blue-luminescent graphene quantum dots." *Advanced Materials* 22, no. 6 (2010): 734–738. doi:10.1002/adma.200902825.

61. Li, Yan, Yue Hu, Yang Zhao, Gaoquan Shi, Lier Deng, Yanbing Hou, and Liangti Qu. "An electrochemical avenue to green-luminescent graphene quantum dots as potential electron-acceptors for photovoltaics." *Advanced Materials* 23, no. 6 (2011): 776–780. doi:10.1002/adma.201003819.

62. Gupta, Vinay, Neeraj Chaudhary, Ritu Srivastava, Gauri Datt Sharma, Ramil Bhardwaj, and Suresh Chand. "Luminescent graphene quantum dots for organic photovoltaic devices." *Journal of the American Chemical Society* 133, no. 26 (2011): 9960–9963. doi:10.1021/ja2036749.

63. Liu, Ruili, Dongqing Wu, Xinliang Feng, and Klaus Müllen. "Bottom-up fabrication of photoluminescent graphene quantum dots with uniform morphology." *Journal of the American Chemical Society* 133, no. 39 (2011): 15221–15223. doi:10.1021/ja204953k.

64. Li, Qiqi, Sheng Zhang, Liming Dai, and Liang-shi Li. "Nitrogen-doped colloidal graphene quantum dots and their size-dependent electrocatalytic activity for the oxygen reduction reaction." *Journal of the American Chemical Society* 134, no. 46 (2012): 18932–18935. doi:10.1021/ja309270h.

65. Tang, Libin, Rongbin Ji, Xiangke Cao, Jingyu Lin, Hongxing Jiang, Xueming Li, Kar Seng Teng et al. "Deep ultraviolet photoluminescence of water-soluble self-passivated graphene quantum dots." *ACS Nano* 6, no. 6 (2012): 5102–5110. doi:10.1021/nn300760g.

66. Dong, Yongqiang, Jingwei Shao, Congqiang Chen, Hao Li, Ruixue Wang, Yuwu Chi, Xiaomei Lin, and Guonan Chen. "Blue luminescent graphene quantum dots and graphene oxide prepared by tuning the carbonization degree of citric acid." *Carbon* 50, no. 12 (2012): 4738–4743. doi:10.1016/j.carbon.2012.06.002.

67. Shen, Chao, Shuyan Ge, Youyou Pang, Fengna Xi, Jiyang Liu, Xiaoping Dong, and Peng Chen. "Facile and scalable preparation of highly luminescent N, S co-doped graphene quantum dots and their application for parallel detection of multiple metal ions." *Journal of Materials Chemistry B* 5, no. 32 (2017): 6593–6600. doi:10.1039/C7TB00506G.

68. Xu, Fengzhou, Hui Shi, Xiaoxiao He, Kemin Wang, Dinggeng He, Xiaosheng Ye, Jinlu Tang, Jingfang Shangguan, and Lan Luo. "Masking agent-free and channel-switch-mode simultaneous sensing of Fe^{3+} and Hg^{2+} using dual-excitation graphene quantum dots." *Analyst* 140, no. 12 (2015): 3925–3928. doi:10.1039/C5AN00468C.

69. Anh, Nguyen Thi Ngoc, Ankan Dutta Chowdhury, and Ruey-an Doong. "Highly sensitive and selective detection of mercury ions using N, S-codoped graphene quantum dots and its paper strip based sensing application in wastewater." *Sensors and Actuators B: Chemical* 252 (2017): 1169–1178. doi:10.1016/j.snb.2017.07.177.

70. Wang, Zhiyu, and ZheFeng Fan. "Cu^{2+} modulated nitrogen-doped grapheme quantum dots as a turn-off/on fluorescence sensor for the selective detection of histidine in biological fluid." *Spectrochimica Acta Part A: Molecular and Biomolecular Spectroscopy* 189 (2018): 195–201. doi:10.1016/j.saa.2017.08.003.

71. Hua, Mengjuan, Chengquan Wang, Jing Qian, Kan Wang, Zhenting Yang, Qian Liu, Hanping Mao, and Kun Wang. "Preparation of graphene quantum dots based core-satellite hybrid spheres and their use as the ratiometric fluorescence probe for visual determination of mercury (II) ions." *Analytica Chimica Acta* 888 (2015): 173–181. doi:10.1016/j.aca.2015.07.042.

72. Sun, Xiangying, Pengchao Liu, Lulu Wu, and Bin Liu. "Graphene-quantum-dots-based ratiometric fluorescent probe for visual detection of copper ion." *Analyst* 140, no. 19 (2015): 6742–6747. doi:10.1039/C5AN01297J.

73. Zhao, Xian-En, Cuihua Lei, Yue Gao, Han Gao, Shuyun Zhu, Xue Yang, Jinmao You, and Hua Wang. "A ratiometric fluorescent nanosensor for the detection of silver ions using graphene quantum dots." *Sensors and Actuators B: Chemical* 253 (2017): 239–246. doi:10.1016/j.snb.2017.06.086.

7 Core–Shell Nanostructures as Plasmonic Nanosensors

7.1 INTRODUCTION

Nanomaterials are those materials that subsequently display innovative possessions from their bulk materials and have at least one or more dimensions in >100 nm scale in the nanometer range. Among the wide-ranging fields of nanotechnology areas, the most important sections are the applications, characterization, and synthesis of NPs. Recently, in the transition's field from microparticles to NPs, NPs have fascinated researchers' attention, leading to enormous variations in a material's chemical and physical possessions. Among various characteristics of NPs, the most important features on a nanoscale are observed. The surface-to-volume ratio increases with the reduction of size of NPs due to the increased number of surface atoms over interior atoms. The small size of the particles corresponds to the area where quantum effects dominate. There is a wide variety of methods accessible for fabricating diverse types of NPs because the development of NPs is a complex process. Resultantly, it is impossible to simplify all of the present existing synthesis methods. However, solid-state processes such as milling, synthesis thru chemical reaction, and condensation from vapor are the essential categories that are broadly used. Based on the suitability of applications such as hydrophobic or hydrophilic content, coated or hybrid NPs can be developed through the above methods. Single NPs have enhanced properties in contrast to bulk materials. Therefore, it has been widely studied by researchers. After that, researchers found in the late 1980s that heterogeneous and colloidal sandwich semiconductor particles have improved competence compared to their single particles [1]. Concentric multilayer semiconductor NPs have been synthesized to improve the property of such semiconductor materials, more recently during the early 1990s. Therefore, the terminology "core/shell" was subsequently approved [2]. Owing to the incredible requirement for additional advanced materials, there has been a regular upsurge in research action fueled through current technology demands. The progression of characterization methods has also been significantly facilitated concurrently to find the structures of these diverse core/shell nanostructures.

DOI: 10.1201/9781003128281-7

7.2   DIFFERENT SHAPED NPs

To develop various shapes nanostructures for example prism [3,4], cube [5,6], octahedron [7,8], hexagon [9,10], wire [11–13], disk [14], tube [15,16], rod [17–19], as well as the spherical (symmetrical) shape NPs, the developments in novel fabrication methods make it probable. As described in some current research, diverse-shaped core/shell NPs are also exceedingly attainable [20–22]. The NPs possessions are linked by the actual shape as well as dependent on their size. For instance, certain properties of magnetic nanocrystals are all dependent on particle size, such as permanent magnetization, magnetic saturation, and blocking temperature. The coercivity depends on the particle shape of the nanocrystals owing to surface anisotropy effects [23]. For example, in the field, high-density information storage diverse-shaped magnetic nanocrystals own great competence in serving our basic considerate of technological applications and magnetism [24]. Other chemical and physical properties of NPs, for example, electrical, selectivity, and catalytic activity [25–27], melting point [28], and optical properties [29,30], are similarly all exceedingly shape-dependent. Other properties of NPs, such as the sensitivity to SERS and plasmon resonance features of silver or AuNPs, also depend on the particle morphology [31].

7.3   CLASSES OF CORE/SHELL NPs

Based on the multiple or single materials, NPs can be categorized into composite or core/shell and simple NPs. Two or more materials are required to synthesize the core/shell and composite particles, whereas single material is used to fabricate the simple NPs. The inner material (core) and an outer layer material (shell) are used to develop the core/shell-type NPs. The inorganic/organic, inorganic/inorganic, organic/organic, and organic/inorganic materials are combined in close interaction to develop these core/shell nanomaterials. The selection of shell material sturdily depends on the use and end application of the core/shell NPs. In spherical core/shell NPs, a simple spherical core particle is entirely coated thru a shell of diverse material in different core–shell NPs. Owing to their diverse, innovative possessions, diverse-shaped core/shell, NPs fascinated by vast research interests. A non-spherical core is utilized to develop diverse-shaped core/shell NPs. When a single shell material is coated onto many small core particles, multiple core/shell particles are produced.

7.4   APPROACHES FOR SYNTHESIS OF CORE/SHELL NP

The "top-down" and "bottom-up" methods are the wide-ranging two categories for nanomaterial synthesis. The "top-down" method outwardly regulated instruments are used to shape, cut, and mill constituents into the required shape and order. It often uses microfabrication methods or traditional workshops. For instance, the most common methods are mechanical techniques (e.g., polishing, grinding, and machining) [32–35], laser-beam processing [36], and lithographic techniques (e.g., ion or electron beam, UV, near optical field, scanning probe) [37–39]. On the other hand, "bottom-up" methods exploit the molecular chemical properties to self-assemble the molecules into some beneficial shapes. Chemical vapor deposition, chemical synthesis, self-assembly, laser tapping (viz., laser-induced assembly), growth, film deposition, and colloidal aggregation are the most common bottom-up approaches [40–42]. The bottom-up and top-down approaches are both not superior currently and have advantages and disadvantages. Though the bottom-up approach can yield tiny-sized particles in contrast to the top-down approach, minimum energy loss, cost-efficiency, and absolute precision are advantages due to their complete control over the process. The bottom-up approach has been established as the utmost technique for synthesizing core/shell NPs to attain an even coating of the shell materials during particle formation. Both approaches can be used as a combination; for example, a bottom-up approach is used to maintain precise and uniform shell thickness, whereas the core particles developed through the top-down approach. A microemulsion is better for using a bulk medium to control the shell thickness and overall size since water droplets act as a nanoreactor or a template.

7.5 APPLICATIONS OF CORE–SHELL NANOSENSORS IN THE SENSING OF HEAVY METALS

Thru observing the color change by naked eyes, heavy metal ions in aqueous media can be conveniently and directly detected. In this addition, a color-dependent operational colorimetric sensor has been synthesized for the facile and selective sensing of Cd^{2+} metal ions based on the of NPs aggregation mechanism. The developed AuNPs capped by carbon quantum dots (CQDs) uniformly further doped with colloidal graphite-like nitride of 15 nm (Au@g-CNQDs). Selectively detect metal by surface functional moieties, that is, hydroxyl, carboxyl, and heptazine groups, which reinforced detection of the Cd^{2+} ions through the "cooperative effect" on Au@g-CNQDs surface. The Cd^{2+} ions could quickly and sensitively capture within 30 s at an ultra-low concentration by the naked eyes without using intricate and expensive special additives or other exogenic indicators. The Cd^{2+} ions sensor owns virtuous analytical presentations within a wide linear range of 0.01–3.0 μM under the optimal conditions, with an LOD up to 10 nM and (S/N=3) [43].

In another study, Pb^{2+} ions in aqueous samples have been detected using a highly sensitive, novel, and straightforward plasmonic colorimetric method. The detection technique is based on the aggregation process in the presence of the Pb^{2+} ions of SiO_2 core–shell nanocomposites (SiO_2@AuNCs) or AuNPs. The SPR peak intensity decreases of SiO_2@AuNCs or AuNPs, which is observed thru adding the Pb^{2+} ions, which also resulted in the color change of SiO_2@AuNCs or AuNPs. Therefore, the Pb^{2+} ions sensing done thru the fabrication of SERS sensitive materials corresponds to the synthesis of nanosensors in the size range of ~360 nm SiO_2@AuNCs and ~20 nm spherical AuNPs. The LSPR peaks at

522 and 541 nm were observed for AuNPs and SiO_2@AuNCs, correspondingly because of their versatility. The UV–vis, TEM, and SEM spectrophotometer characterize the developed AuNPs and SiO_2@AuNCs with or without the analyte concentration. The SiO_2@Au NCs with the size 360 nm show the detection limit of 50 nM for Pb^{2+} ions, whereas AuNPs of 20 nm size have the detection limit of 500 nM. The result showed that for sensing Pb^{2+} ions, ten times more sensitivity had been shown by SiO_2@AuNCs in contrast to AuNPs [44].

Similarly, for the removal and detection of Ag^+ and Hg^{2+} ions selectively, a SERS active platform was developed through the magnetic silica sphere functionalized with oligonucleotide NPs (MSS)@AuNPs. The developed probe exploits mismatched C–Ag–C and T–Hg–T bridges, demonstrating outstanding responses for Ag^+ and Hg^{2+} ions in the range of 10–1,000 and 0.1–1,000 nM, correspondingly. By spontaneous precipitation or thru an exterior magnetic field, the developed probe can effectively remove the Ag^+ and Hg^{2+} ions from surrounding solutions. Furthermore, using cysteine, MSS@AuNPs can simply cast off over 80% [45].

In addition, in aqueous media, using silica cores (420 nm diameter) and spherical gold AuNPs (30 nm diameter), the silica-gold core–shell (SiO_2@Au) nanocomposites have been synthesized, which was further used for Cd^{2+} ion detection at their trace level. The SPR peak of synthesized core–shell particles was observed at 541 and 522 nm for AuNPs. In aqueous samples, for the sensing of Cd^{2+} ions, both the developed SiO_2@Au and AuNPs solutions showed good selectivity. The AuNPs and SiO_2@Au show the LOD at a concentration of Cd^{2+} ions at 2 and 0.1 ppm, correspondingly. The results showed that the developed SERS active SiO_2@Au core–shell NCs showed 20 times more sensitivity for the detection method than the AuNPs solution. In the existence of Cd^{2+} ions, the SiO_2@Au particles and AuNPs showed aggregation, which can be observed by the SEM analysis [46].

Similarly, an economical, rapid, and facile colorimetric method for sensing Hg^{2+} ions has been developed using marine mussel (3,4-dihydroxyphenylalanine (DOPA)), which is inspired thru its redox property and bioadhesive property. The presence of ample catechol chains would reduce Hg^{2+} to Hg^0 on the Ag@DOPA surface, and with the Hg^0 deposition, it forms Ag@DOPA@Hg nanostructure. The XPS, TEM, elemental maps, and FTIR were used to characterize the nanostructure of Ag@DOPA before and after the addition of Hg^{2+} ions. Due to the formation of Ag@DOPA@Hg nanostructure after adding Hg^{2+} ions, the alteration in color from golden yellow to purple-blue of Ag@DOPA occurs thru the colorimetric method within 5 min. The SPR spectral band of Ag@DOPA colloid showed a virtuous linear relationship, and with increasing Hg^{2+} ion concentrations from 10 nM to 4 µM, the SPR of Ag@DOPA decreased. The developed probe showed the LOD at 5 nM [47].

In another study, silver nanoshells have been developed using the chemical reduction method to sense Hg^{2+} ions in an aqueous solution. The SPR peak of silver nanoshells was observed at 450 nm, whereas silver NPs display a strong SPR peak at 433 nm. The developed probe showed the detection limit for Hg^{2+}

ions found to be 0.01 mL with good sensitivity in the water. The silver nanoshells of 202 nm showed the detection limit of 28 nm shell thickness for Hg^{2+} ions. The shapes and sizes of nanoshells can be altered by choosing the capping and reducing agents. TEM and UV–vis spectroscopy were used for the characterization of the NPs [48].

For detecting Hg^{2+} ions, a cost-effective and simple colorimetric method has been synthesized using unmodified Au@Ag nanorods (NRs). The etching sensing mechanism among the Hg^{2+} ions and Ag nanoshell of Au@Ag NRs was associated with the redox reaction. An alteration in color occurs from brownish red to light red after adding the Hg^{2+} ion to the Au@AgNRs solution, which further turned into light-violet and finally to colorless. The developed method showed the detection limit of 10 nM for Hg^{2+}. Using Au@AgNRs was utilized with a thick Ag nanoshell, a dip in thickness was detected located among two strong absorption peaks. A new sensor parameter has been attained as the alteration in this dip for the Hg^{2+} ion detection based on absorption spectra. In a mixed metal ion solution, the developed probe selectively detects Hg^{2+} ions [49].

Similarly, for detecting metal ions, a detection probe has been developed as DNA scaffold Ag–Au alloy nanoclusters (Ag–Au ANCs). The metal nanoclusters used to owe to their great application potentials and unique properties. The Ag–Au ANCs have been synthesized using a one-pot wet-chemical strategy. Various analytical tools have been used to characterize the developed nanoclusters, i.e., mass spectrometry (MS), XPS, and TEM. Here, the AgNCs act as core, and Au behaves as shell material, which resulted from the strong interaction between Ag^+ and DNA. Compared to AgNCs, the fluorescence of Ag–Au ANCs was discovered for the detection of Hg^{2+}. The Ag–Au ANCs showed selective detection of Hg^{2+} ions with excellent and highly improved linearity even in the presence of Cu^{2+} ion solution [50].

In another study, Fe_3O_4@Au magnetic NPs changed with DNA have been developed as an electrochemical biosensor. In this, to comprise certain mismatched base pairs, three DNA probes are designed. The first probe named DNA 1 is modified and thiolated on Fe_3O_4@Au NPs surface, and DNA 2 and DNA 3 are labeled with two independent electrochemical species. The DNA 1/DNA 3 hybridization and DNA 1/DNA 2 hybridization can be assisted by stable structures of thymine–Hg^{2+}–thymine and cytosine–Ag^+–cytosine formed in the existence of Hg^{2+} and Ag^+, which detect consistent electrochemical moieties on the surface of magnetic NPs. Through recording the square wave voltammetry signals, the accomplished nanocomposites were utilized as electrochemical detection probes for selective sensing of metals. The developed probe selectively detects Hg^{2+} and Ag^+ without significant interference [51]. Figure 7.1 shows the SEM images of core–shell NPs before and after sensing of metal ions.

In addition, cellulose nanocrystal–alginate hydrogel beads loaded bovine serum albumin-protected with gold nanoclusters (Au@BSA NCs) developed for the heavy metal ion detection precisely Hg^{2+} ions in water. Various analytical tools were utilized to characterize the detection system, i.e., XPS and SEM, coupled with EDX for predicting the chemical and physical characteristics of the

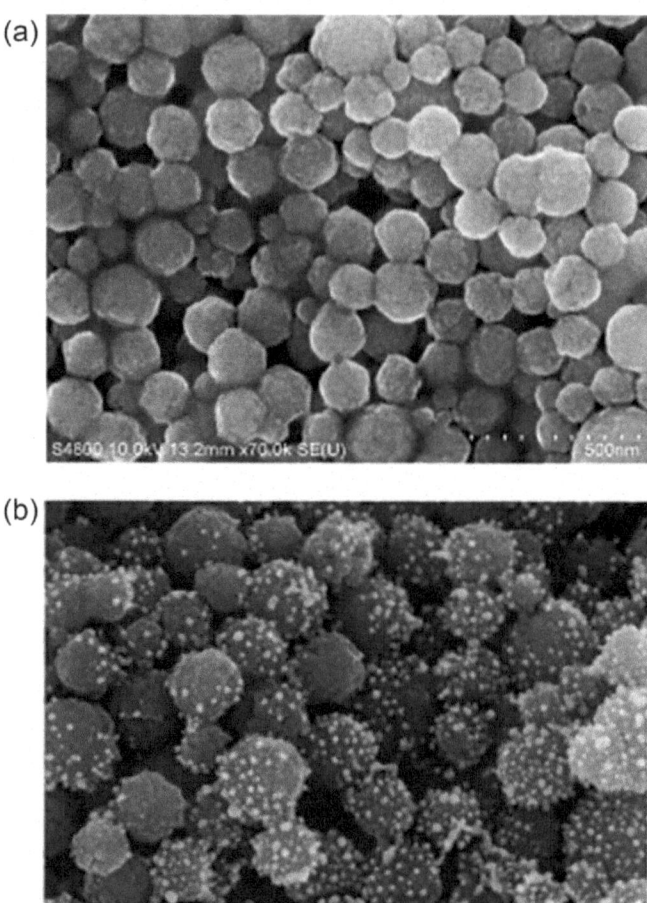

FIGURE 7.1   SEM images of (a) bare NPs and (b) core–shell NPs. (Reprinted with permission from Ref. [51]. Copyright 2017 American Chemical Society.)

system. Upon binding with Hg^{2+} ions, a fluorescence-quenching dynamic of Au@ BSA NCs observed shows that a diffusion phenomenon developed using hydrogel beads [52].

Similarly, for detecting and removing Hg^{2+} contamination, ZnO/Ag nanoarrays active toward SERS have been developed. The developed probe detects toxic Hg^{2+} ions when it is treated with heat and detects when exposed to UV from the SERS marker, consequently the capability of self-cleaning of this photocatalyst. Among other chemicals, the sensors are highly selective due to the inimitable way of Hg^{2+} ions, thus reducing SERS activity through interaction with AgNPs [53]. Figure 7.2 shows the schematic representation of the detection mechanism for heavy metal ions.

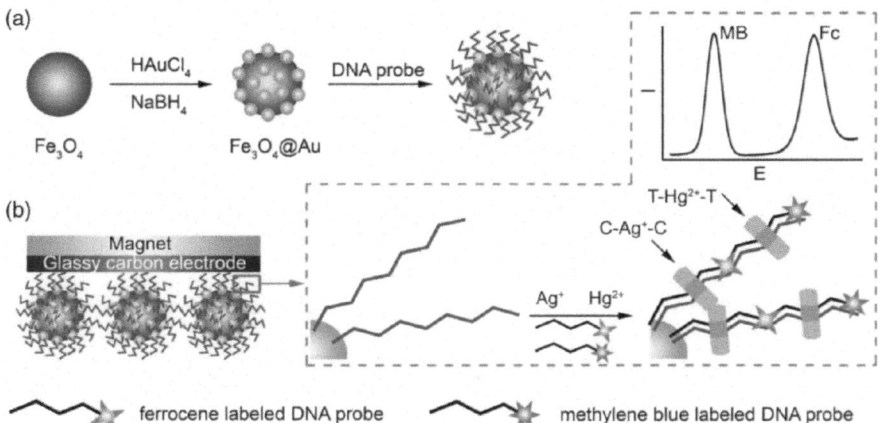

FIGURE 7.2 Schematic representation of detection mechanism for heavy metal ions. (Reprinted with permission from Ref. [51]. Copyright 2017 American Chemical Society.)

Similarly, a new NP-containing test paper sensor has been developed to detect Cu^{2+} ions in aqueous solutions that can be utilized as an easy to use, inexpensive, highly selective, and portable sensor. For the fabrication of this disposable paper sensor, the core–shell NPs of ZnO@ZnS have been utilized. The ZnO@ZnS was used for coating the paper, and a chemical method was utilized for the synthesis process. The optimum temperature for the fabrication of the ZnO@ZnS core–shell NPs was 60 °C, which is the lowest range of temperature. The developed probe showed the LOD as low at 0.96 ppm (~15 μM) for different concentrations of aqueous Cu^{2+} ions. The detection of Cu^{2+} ions concentration with high selectivity shows a maximum 5% relative error using the paper-based sensor. The developed probe can use the employed disposable paper-based sensor to synthesize a selective, precise, and portable detection method for heavy metals [54].

In another study, to sense the aqueous Hg^{2+} ions, the AuNPs supported porous MoS_2 composite aerogel has been fabricated. The developed porous $Au/Fe_3O_4/MoS_2CAs$ aerogel showed a low detection limit up to 3.279 nM for sensitively detect Hg^{2+} ions in an aqueous solution. The developed colorimetric technique also showed the adsorption of Hg^{2+} ions with an adsorption capacity up to 1,527 mg g^{-1} and rapid desorption ability. Within a few minutes, the Hg^{2+} ions levels diminished from 10 ppm to 0.11 ppb up to below 2 ppb after magnetic separation. $Au/Fe_3O_4/MoS_2CAs$ showed high removal efficiency (RF) up to >95% and could be sequentially recycled more than ten times. The strong coupling among MoS_2 nanosheets and AuNPs the developed aerogel showed excellent performance from its 3D interconnected macroporous framework, interpreting it as a capable adsorbent and sensing material for environmental remediation toward Hg^{2+} ions contaminated water [55].

Through $FeCl_3$ oxidative coupling polymerization, a carbazolic porous framework (Cz-TPM) with a tetrahedral core was synthesized. The Brunauer–Emmett–Teller (BET) surface area of the obtained polymers was $713.2\,m^{-2}g^{-1}$. Moreover, the developed Cz-TPM showed outstanding capability toward the sensing of mercury and iodide ions. After the addition of iodide and addition of Hg^{2+} salts, the Cz-TPM dispersion turns the color of the solution from yellow to colorless. The Hg^{2+} ions can recognize through "turn on" fluorescence recovery using in situ produced Cz-TPM@I complexes, and I^- can be detected thru "turn off" fluorescence quenching with produced Cz-TPM [56].

In a similar study, for the sensing of Hg^{2+} ions, Fe_3O_4@$nSiO_2$@$mSiO_2$-TTRDNA nanocomposite has been fabricated. The thymine (T) and T-rich DNA functionalization were done to synthesize highly selective and regenerable Fe_3O_4@$nSiO_2$@$mSiO_2$ nanocomposite. Here, the thymine and T-rich DNA were immobilized onto the interior and exterior surface of the outermost mesoporous silica, correspondingly. The appearance of strong fluorescence by the addition of SYBR Green I dye shows the detection mechanism. Based on Hg^{2+}- mediated hairpin structure, the mechanism occurs thru the functionalization of T-rich DNA on the exterior surface of the nanocomposites. Though, in the absence of Hg^{2+} ions, low fluorescence results from the addition of the dye. The developed probe showed the LOD by fluorescence spectroscopy at 2 nM for Hg^{2+} ions in a buffer. Concurrently, together with the selective sensing of Hg^{2+} ions, the developed Fe_3O_4@$nSiO_2$@$mSiO_2$-T-TRDNA nanocomposite also showed convenient and efficient elimination of Hg^{2+} ion by a magnet. The kinetic data showed that within around 1 h, the synthesized probe divulges rapid removal of Hg^{2+} over 80%. Using resistance to nuclease digestion and simple acid treatment, the unique property of the developed nanocomposites also includes the regeneration process for the removal and detection of Hg^{2+}. Similar processes can be used with other small molecules and nucleic acids to functionalize the Fe_3O_4@$nSiO_2$@$mSiO_2$ nanocomposites [57].

In aqueous systems for sensing Hg^{2+} ions, a colorimetric "turn-on" fluorescent chemosensor has been developed, that is, Rho-Fe_3O_4@SiO_2. In this, Fe_3O_4@SiO_2 magnetic core–shell NPs are conjugated with the N-(rhodamine-6G)lactamethylenediamine (Rho-en) has been synthesized and designed. Various analytical tools such as XRD, TEM, fluorescence emission, and UV–vis and FTIR spectra were used to characterize the developed probe. The multifunctional NPs Rho-Fe_3O_4@SiO_2 has shown the selective "turn-on"-type fluorescent enhancements, which were further verified by UV–visible spectra and fluorescence together with distinct color changes by the addition of Hg^{2+} ions. In the same conditions, in contrast to Rho-en, the Rho-Fe_3O_4@SiO_2 showed better selectivity for Hg^{2+} ions. In the sensor, magnetic Fe_3O_4 NPs facilitated the magnetic separation of the Hg(II)–Rho-Fe_3O_4@SiO_2 from the solution [58].

In another study, to synthesize manageable core–shell NPs, a novel shell method is used. The core interacts with metal ions to form the shell, and by the use of $Na_2S_2O_3$, the thickness and size of the shell and the core can be regulated, respectively. The shell is made of complex insoluble salts containing Ag_3AuS_2,

AuAgS, and Au_2S by analyzing TEM and XRD. Ag_3AuS_2, AuAgS, and Au_2S shell development occur from the exterior part, resulting in the ensuing core–shell NPs attained at diverse reaction centers. In aqueous media, to sense Ag^+ ions, the developed core–shell NPs can be utilized as nanosensors fabricated thru this method with high sensitivity and selectivity. In mixed metal ion solution, the developed probe showed excellent selectivity for Ag^+ ions. Moreover, the research showed that GNRs offer higher sensitivity than gold nanospheres [59].

Similarly, the Fe_3O_4@C@CdTe fluorescent core/shell microspheres have been fabricated by the layer-by-layer (LBL) self-assembly method and hydrothermal method. This innovative nanosensor was synthesized by developing a magnetic core (Fe_3O_4) with a shell made of CdTe fluorescent QDs. For ultrasensitive detection of Cu^{2+} ion, the Fe_3O_4@C@CdTe fluorescent core/shell microspheres are utilized as a chemosensor. The fluorescence quenched efficiently thru Cu^{2+} ions of the attained chemosensor. The presence of static and dynamic quenching methods was observed by examining the quenching mechanism, demonstrating that static quenching is more protuberant out of both processes. The $110\,\mu M$, a quenching constant (K_{sv}) of 4.9×10^{-4} M1 is shown by the changed Stern–Volmer equation with a good linear response ($R^2 = 0.9957$). Primarily, the Fe_3O_4@C@CdTe core/shell microspheres could be collected and separated simply by a commercial magnet in 10 s and due to its superparamagnetic nature. These outcomes attained the development of other nanostructure-based multifunctional hybrid nanocomposites and offered a method to resolve the discomfitures in applied detection applications of QDs [60].

In a similar study, (Fe_3O_4@$Mn_xFe_yO_4$) core–shell nanocomposites have been developed for low-field and ultra-high measurement of uranium ions through the surface-optimized method in water. The developed nanocomposite is surface stabilized through an (organic) phosphate functionalized bilayer to form multifunctional (Fe_3O_4@$Mn_xFe_yO_4$) core–shell nanocomposites [61].

Using many detection methods and materials, chemical detection for the suitable sensing of Hg^{2+} ions has been extensively discovered. It remains a challenge to attain sensitive but inexpensive, fast, and simple recognition of toxic metals. SERS chip has been developed to detect Hg^{2+} ions at the femtomolar (fM) level. The probe was developed using an organic ligand (4,4′-Dipyridyl (Dpy)) with Au@Ag NPs. 4,4′-Dipyridyl (Dpy) was utilized to produce robust SERS readouts and Raman hot spots. Through its dual interacting sites to Ag nanoshells, it also can control the aggregation of Au@AgNPs. However, the SERS signals of Dpy are quenched through the coordination of Hg^{2+} ions, which can hinder the aggregation of Au@AgNPs. A SERS chip has been developed based on these findings thru further modification with Dpy on a piece of silicon wafer followed by the assembly of Au@Ag NPs. Each test accomplished within ~4 min using the SERS chip and is only needed a droplet of $20\,\mu L$ sample. The SERS chip has also been utilized to sensing Hg^{2+} in lake water, juice, and milk [62].

Similarly, due to the worldwide demand for portable, sensitive, cost-effective, and user-friendly biosensors, a biosensor based on a barometer has been developed to recognize a wide-ranging analyte. Using a portable barometer in a limited chamber, the change in pressure produced by the generation of oxygen is measured by the device. To catalyze the release of O_2 and the decomposition of H_2O_2, the Au@Pt core–shell NPs fabricated as the bioassay detection probe. Barometer-based immune sensors were developed as a standard probe to detect ractopamine (Rac) and carcinoembryonic antigen (CEA). For sensitive recognition of mercury ion (Hg^{2+}) and thrombin, barometer-based aptasensors were developed. To wirelessly transmit, save, and calculate, smartphone software has been fabricated to increase the result analysis. The established probe showed the detection process in a linear range for Rac, CEA, Hg^{2+}, and thrombin were 0.0625–4, 0.025–1.6, 0.25–16 ng mL^{-1}. and 4–128 U L^{-1}, respectively. The four analytes have the detection limit of 0.051, 0.021, 0.22 and 2.4 U L^{-1}, correspondingly. For these four analytes, the established barometer-based biosensors showed high specificities. The barometer-based biosensors measured the Hg^{2+} in river water samples, CEA in serum samples, Rac in urine samples, and thrombin in serum samples. In contrast to traditional methods, barometer-based biosensors are widely applicable due to their consistency with sensing outcomes toward target analytes [63].

7.6 CONCLUSION AND FUTURE PERSPECTIVES

This chapter focuses on various classes of core/shell NPs for sensing applications. Future generations of core/shell NPs will exhibit various novel possessions, which will result in innovative benefits with enhanced competence. Usually, for being capable of shielding the core material from the neighboring atmosphere, for enhanced steadiness, for easy biofunctionalization, enhanced chemical and physical properties, and enhanced semiconductor possessions, the core/shell NPs are well-known candidates. Because of their extensive usage in the electronics and biomedical fields, diverse core/shell NPs are widely studied. For example, fluorescence and magnetic core materials coated with other inorganic materials and inorganic/inorganic NPs. Some organic/inorganic and organic/organic core–shell NPs must be essential to increase their use in different industrial fields. Due to variations in some of their chemical properties, that is, oxidation stability, and physical properties that are abrasion capability, mechanical strength, and glass transition temperature, some organic/inorganic and organic/organic core–shell NPs must be essential to increase their use in different industrial fields. Core/shell NPs are very important from the application point of view also, a wide range of applications such as electronics, biomedical applications, catalysis, and biomedical and chemical sensors. A mechanistic insight will be needed to develop innovative mechanisms to facilitate the fabrication method and synthesize novel core/shell materials where the future is concerned. From laboratory to clinical trials corresponding to novel application areas, biological applications are predictable to be advanced. The significance of method scalability should not be undervalued from a laboratory to a commercial scale. In conclusion, it is likewise expected that additional theoretic studies will correspond to facilitate experimental trials and improve consideration with such advances in the experimental work. It is expected that in serving future developments in chemical sensing of heavy metal ions using the core/shell-based nanosensors, this chapter will play a key role for the readers.

REFERENCES

1. Wei, Xian-Wen, Guo-Xing Zhu, Yuan-Jun Liu, Yong-Hong Ni, You Song, and Zheng Xu. "Large-scale controlled synthesis of FeCo nanocubes and microcages by wet chemistry." *Chemistry of Materials* 20, no. 19 (2008): 6248–6253. doi:10.1021/cm800518x.

2. Wu, Yan, Peng Jiang, Ming Jiang, Tie-Wei Wang, Chuan-Fei Guo, Si-Shen Xie, and Zhong-Lin Wang. "The shape evolution of gold seeds and gold@ silver core–shell nanostructures." *Nanotechnology* 20, no. 30 (2009): 305602. doi:10.1088/0957-4484/20/30/305602.

3. Hu, Jin-Song, Yu-Guo Guo, Han-Pu Liang, Li-Jun Wan, and Li Jiang. "Three-dimensional self-organization of supramolecular self-assembled porphyrin hollow hexagonal nanoprisms." *Journal of the American Chemical Society* 127, no. 48 (2005): 17090–17095. doi:10.1021/ja0553912.

4. Jitianu, Mihaela, and Dan V. Goia. "Zinc oxide colloids with controlled size, shape, and structure." *Journal of Colloid and Interface Science* 309, no. 1 (2007): 78–85. doi:10.1016/j.jcis.2006.12.020.

5. Salazar-Alvarez, G., Jian Qin, V. Sepelak, I. Bergmann, M. Vasilakaki, K. N. Trohidou, J. D. Ardisson et al. "Cubic versus spherical magnetic nanoparticles: The role of surface anisotropy." *Journal of the American Chemical Society* 130, no. 40 (2008): 13234–13239. doi:10.1021/ja0768744.

6. Chaubey, Girija S., Carlos Barcena, Narayan Poudyal, Chuanbing Rong, Jinming Gao, Shouheng Sun, and J. Ping Liu. "Synthesis and stabilization of FeCo nanoparticles." *Journal of the American Chemical Society* 129, no. 23 (2007): 7214–7215. doi:10.1021/ja0708969.

7. Schmidt, Erik, Angelo Vargas, Tamas Mallat, and Alfons Baiker. "Shape-selective enantioselective hydrogenation on Pt nanoparticles." *Journal of the American Chemical Society* 131, no. 34 (2009): 12358–12367. doi:10.1021/ja9043328.

8. Wang, Z. L., T. S. Ahmad, and M. A. El-Sayed. "Steps, ledges and kinks on the surfaces of platinum nanoparticles of different shapes." *Surface Science* 380, no. 2–3 (1997): 302–310. doi:10.1016/S0039-6028(97)05180-7.

9. Ahmed, Jahangeer, Shudhanshu Sharma, Kandalam V. Ramanujachary, Samuel E. Lofland, and Ashok K. Ganguli. "Microemulsion-mediated synthesis of cobalt (pure fcc and hexagonal phases) and cobalt–nickel alloy nanoparticles." *Journal of Colloid and Interface Science* 336, no. 2 (2009): 814–819. doi:10.1016/j.jcis.2009.04.062.

10. El-Safty, Sherif A. "Synthesis, characterization and catalytic activity of highly ordered hexagonal and cubic composite monoliths." *Journal of Colloid and Interface Science* 319, no. 2 (2008): 477–488. doi:10.1016/j.jcis.2007.12.010.

11. Cao, Guozhong, and Dawei Liu. "Template-based synthesis of nanorod, nanowire, and nanotube arrays." *Advances in Colloid and Interface Science* 136, no. 1–2 (2008): 45–64. doi:10.1016/j.cis.2007.07.003.

12. Wu, Yiying, Tsachi Livneh, You Xiang Zhang, Guosheng Cheng, Jianfang Wang, Jing Tang, Martin Moskovits, and Galen D. Stucky. "Templated synthesis of highly ordered mesostructured nanowires and nanowire arrays." *Nano Letters* 4, no. 12 (2004): 2337–2342. doi:10.1021/nl048653r.

13. Liu, Zhu, David Elbert, Chia-Ling Chien, and Peter C. Searson. "FIB/TEM characterization of the composition and structure of core/shell Cu– Ni nanowires." *Nano Letters* 8, no. 8 (2008): 2166–2170. doi:10.1021/nl080492u.

14. Qu, Xiaozhong, Leila Omar, Thi Bich Hang Le, Laurence Tetley, Katherine Bolton, Kar Wai Chooi, Wei Wang, and Ijeoma F. Uchegbu. "Polymeric amphiphile branching leads to rare nanodisc shaped planar self-assemblies." *Langmuir* 24, no. 18 (2008): 9997–10004. doi:10.1021/la8007848.

15. Shin, Tae-Yeon, Sang-Hoon Yoo, and Sungho Park. "Gold nanotubes with a nanoporous wall: Their ultrathin platinum coating and superior electrocatalytic activity toward methanol oxidation." *Chemistry of Materials* 20, no. 17 (2008): 5682–5686. doi:10.1021/cm800859k.

16. Xiao, Rui, Seung Il Cho, Ran Liu, and Sang Bok Lee. "Controlled electrochemical synthesis of conductive polymer nanotube structures." *Journal of the American Chemical Society* 129, no. 14 (2007): 4483–4489. doi:10.1021/ja068924v.

17. Park, Sungho, Jung-Hyurk Lim, Sung-Wook Chung, and Chad A. Mirkin. "Self-assembly of mesoscopic metal-polymer amphiphiles." *Science* 303, no. 5656 (2004): 348–351. doi:10.1126/science.1093276.

18. Peng, Xiaogang, Liberato Manna, Weidong Yang, Juanita Wickham, Erik Scher, Andreas Kadavanich, and A. Paul Alivisatos. "Shape control of CdSe nanocrystals." *Nature* 404, no. 6773 (2000): 59–61. doi:10.1038/35003535.

19. Park, Sungho, Sung-Wook Chung, and Chad A. Mirkin. "Hybrid Organic–Inorganic, Rod-Shaped Nanoresistors and Diodes." *Journal of the American Chemical Society* 126, no. 38 (2004): 11772–11773. doi:10.1021/ja046077v.

20. Fakhraian, Hossein, and Fardin Valizadeh. "Activation of hydrogen peroxide via bicarbonate, sulfate, phosphate and urea in the oxidation of methyl phenyl sulfide." *Journal of Molecular Catalysis A: Chemical* 333, no. 1–2 (2010): 69–72. doi:10.1016/j.molcata.2010.09.017.

21. Radi, Abdullah, Debabrata Pradhan, Youngku Sohn, and K. T. Leung. "Nanoscale shape and size control of cubic, cuboctahedral, and octahedral Cu–Cu$_2$O core–shell nanoparticles on Si (100) by one-step, template less, capping-agent-free electrodeposition." *ACS Nano* 4, no. 3 (2010): 1553–1560. doi:10.1021/nn100023h.

22. Libor, Zsuzsanna, and Qi Zhang. "The synthesis of nickel nanoparticles with controlled morphology and SiO$_2$/Ni core-shell structures." *Materials Chemistry and Physics* 114, no. 2–3 (2009): 902–907. doi:10.1016/j.matchemphys.2008.10.068.

23. Chen, Jingyi, Thurston Herricks, and Younan Xia. "Polyol synthesis of platinum nanostructures: Control of morphology through the manipulation of reduction kinetics." *Angewandte Chemie International Edition* 44, no. 17 (2005): 2589–2592. doi:10.1002/anie.200462668.

24. Song, Qing, and Z. John Zhang. "Shape control and associated magnetic properties of spinel cobalt ferrite nanocrystals." *Journal of the American Chemical Society* 126, no. 19 (2004): 6164–6168. doi:10.1021/ja049931r.

25. Zheng, Yuanhui, Yao Cheng, Yuansheng Wang, Feng Bao, Lihua Zhou, Xiaofeng Wei, Yingying Zhang, and Qi Zheng. "Quasicubic α-Fe$_2$O$_3$ nanoparticles with excellent catalytic performance." *The Journal of Physical Chemistry B* 110, no. 7 (2006): 3093–3097. doi:10.1021/jp056617q.

26. Narayanan, Radha, and Mostafa A. El-Sayed. "Changing catalytic activity during colloidal platinum nanocatalysis due to shape changes: Electron-transfer reaction." *Journal of the American Chemical Society* 126, no. 23 (2004): 7194–7195. doi:10.1021/ja0486061.

27. Lee, Hyunjoo, Susan E. Habas, Sasha Kweskin, Derek Butcher, Gabor A. Somorjai, and Peidong Yang. "Morphological control of catalytically active platinum nanocrystals." *Angewandte Chemie International Edition* 45, no. 46 (2006): 7824–7828. doi:10.1002/anie.200603068.

28. Gupta, Sanjeev K., Mina Talati, and Prafulla K. Jha. "Shape and size dependent melting point temperature of nanoparticles." In *Materials Science Forum*, vol. 570, pp. 132–137. Trans Tech Publications Ltd, 2008. doi:10.4028/www.scientific.net/MSF.570.132.

29. Stuart, D. A., A. J. Haes, C. R. Yonzon, E. M. Hicks, and R. P. Van Duyne. "Biological applications of localised surface plasmonic phenomenae." In *IEE Proceedings-Nanobiotechnology*, vol. 152, no. 1, pp. 13–32. IET Digital Library, 2005. doi:10.1049/ip-nbt:20045012.

30. Millstone, Jill E., Sarah J. Hurst, Gabriella S. Métraux, Joshua I. Cutler, and Chad A. Mirkin. "Colloidal gold and silver triangular nanoprisms." *Small* 5, no. 6 (2009): 646–664. doi:10.1002/smll.200801480.

31. Sun, Yugang, and Younan Xia. "Triangular nanoplates of silver: Synthesis, characterization, and use as sacrificial templates for generating triangular nanorings of gold." *Advanced Materials* 15, no. 9 (2003): 695–699. doi:10.1002/adma.200304652.

32. Dodd, Aaron C. "A comparison of mechanochemical methods for the synthesis of nanoparticulate nickel oxide." *Powder Technology* 196, no. 1 (2009): 30–35. doi:10.1016/j.powtec.2009.06.014.

33. Deng, W. J., W. Xia, C. Li, and Y. Tang. "Formation of ultra-fine grained materials by machining and the characteristics of the deformation fields." *Journal of Materials Processing Technology* 209, no. 9 (2009): 4521–4526. doi:10.1016/j.jmatprotec.2008.10.043.

34. Salari, M., P. Marashi, and M. Rezaee. "Synthesis of TiO_2 nanoparticles via a novel mechanochemical method." *Journal of Alloys and Compounds* 469, no. 1–2 (2009): 386–390. doi:10.1016/j.jallcom.2008.01.112.

35. Sasikumar, R., and R. M. Arunachalam. "Synthesis of nanostructured aluminium matrix composite (AMC) through machining." *Materials Letters* 63, no. 28 (2009): 2426–2428. doi:10.1016/j.matlet.2009.08.017.

36. Hong, L. I., R. M. Vilar, and W. A. N. G. Youming. "Laser beam processing of a SiC particulate reinforced 6061 aluminium metal matrix composite." *Journal of Materials Science* 32, no. 20 (1997): 5545–5550. doi:10.1023/A:1018668322943.

37. Subramanian, R., P. E. Denney, J. Singh, and M. Otooni. "A novel technique for synthesis of silver nanoparticles by laser-liquid interaction." *Journal of Materials Science* 33, no. 13 (1998): 3471–3477. doi:10.1023/A:1013270305004.

38. Kumar, P., Ravi Kumar, D. Kanjilal, M. Knobel, P. Thakur, and K. H. Chae. "Ion beam synthesis of Ni nanoparticles embedded in quartz." *Journal of Vacuum Science & Technology B: Microelectronics and Nanometer Structures Processing, Measurement, and Phenomena* 26, no. 4 (2008): L36–L40. doi:10.1116/1.2956624.

39. Jang, Deoksuk, Bukuk Oh, and Dongsik Kim. "Generation of metal nanoparticles by laser ablation of metal microparticles and plume dynamics." In *High-Power Laser Ablation IV*, vol. 4760, pp. 1024–1031. International Society for Optics and Photonics, 2002. doi:10.1117/12.482061.

40. Sneh, Ofer, Robert B. Clark-Phelps, Ana R. Londergan, Jereld Winkler, and Thomas E. Seidel. "Thin film atomic layer deposition equipment for semiconductor processing." *Thin Solid Films* 402, no. 1–2 (2002): 248–261. doi:10.1016/S0040-6090(01)01678-9.

41. Wang, Y. Y., K. F. Cai, and X. Yao. "Facile synthesis of PbTe nanoparticles and thin films in alkaline aqueous solution at room temperature." *Journal of Solid State Chemistry* 182, no. 12 (2009): 3383–3386. doi:10.1016/j.jssc.2009.10.004.

42. Yoo, Sang-Hoon, Lichun Liu, and Sungho Park. "Nanoparticle films as a conducting layer for anodic aluminum oxide template-assisted nanorod synthesis." *Journal of Colloid and Interface Science* 339, no. 1 (2009): 183–186. doi:10.1016/j.jcis.2009.07.049.

43. Zhang, Zhuoyue, Zhen Zhang, Huihui Liu, Xiang Mao, Wei Liu, Shouting Zhang, Zongxiu Nie, and Xiaoquan Lu. "Ultratrace and robust visual sensor of Cd^{2+} ions based on the size-dependent optical properties of Au@ g-CNQDs nanoparticles in mice models." *Biosensors and Bioelectronics* 103 (2018): 87–93. doi:10.1016/j.bios.2017.12.025.

44. Thatai, Sheenam, Parul Khurana, Surendra Prasad, Sarvesh K. Soni, and Dinesh Kumar. "Trace colorimetric detection of Pb^{2+} using plasmonic gold nanoparticles and silica–gold nanocomposites." *Microchemical Journal* 124 (2016): 104–110. doi:10.1016/j.microc.2015.07.006.

45. Liu, Min, Zhuyuan Wang, Shenfei Zong, Hui Chen, Dan Zhu, Lei Wu, Guohua Hu, and Yiping Cui. "SERS detection and removal of mercury (II)/silver (I) using oligonucleotide-functionalized core/shell magnetic silica sphere@ Au nanoparticles." *ACS Applied Materials & Interfaces* 6, no. 10 (2014): 7371–7379. doi:10.1021/am5006282.

46. Thatai, Sheenam, Parul Khurana, Surendra Prasad, and Dinesh Kumar. "Plasmonic detection of Cd^{2+} ions using surface-enhanced Raman scattering active core–shell nanocomposite." *Talanta* 134 (2015): 568–575. doi:10.1016/j.talanta.2014.11.024.

47. Hu, Yuling, Dongmei Wang, and Gongke Li. "Mussel inspired redox surface for one step visual and colorimetric detection of Hg^{2+} during the formation of Ag@ DOPA@ Hg nanoparticles." *Analytical Methods* 7, no. 15 (2015): 6103–6108. doi:10.1039/C5AY01272D.

48. Boken, Jyoti, and Dinesh Kumar. "Detection of toxic metal ions in water using SiO$_2$@ Ag Core-Shell nanoparticles." *International Journal of Environmental Research and Development* 4 (2014): 303–308.

49. Yang, Rong, Dan Song, Chongwen Wang, Anna Zhu, Rui Xiao, Jingquan Liu, and Feng Long. "Etching of unmodified Au@ Ag nanorods: A tunable colorimetric visualization for the rapid and high selective detection of Hg^{2+}." *RSC Advances* 5, no. 124 (2015): 102542–102549. doi:10.1039/C5RA19627B.

50. Zhang, Tianxiang, Hongwei Xu, Shihan Xu, Biao Dong, Zhongyang Wu, Xinran Zhang, Lihang Zhang, and Hongwei Song. "DNA stabilized Ag–Au alloy nanoclusters and their application as sensing probes for mercury ions." *RSC Advances* 6, no. 57 (2016): 51609–51618. doi:10.1039/C6RA07563K.

51. Miao, Peng, Yuguo Tang, and Lei Wang. "DNA modified Fe$_3$O$_4$@ Au magnetic nanoparticles as selective probes for simultaneous detection of heavy metal ions." *ACS Applied Materials & Interfaces* 9, no. 4 (2017): 3940–3947. doi:10.1021/acsami.6b14247.

52. Mohammed, Nishil, Avijit Baidya, Vasanthanarayan Murugesan, Avula Anil Kumar, Mohd Azhardin Ganayee, Jyoti Sarita Mohanty, Kam Chiu Tam, and Thalappil Pradeep. "Diffusion-controlled simultaneous sensing and scavenging of heavy metal ions in water using atomically precise cluster–cellulose nanocrystal composites." *ACS Sustainable Chemistry & Engineering* 4, no. 11 (2016): 6167–6176. doi:10.1021/acssuschemeng.6b01674.

53. Esmaielzadeh Kandjani, Ahmad, Ylias M. Sabri, Mahsa Mohammad-Taheri, Vipul Bansal, and Suresh K. Bhargava. "Detect, remove and reuse: A new paradigm in sensing and removal of Hg (II) from wastewater via SERS-active ZnO/Ag nanoarrays." *Environmental Science & Technology* 49, no. 3 (2015): 1578–1584. doi:10.1021/es503527e.

54. Sadollahkhani, Azar, Amir Hatamie, Omer Nur, Magnus Willander, Behrooz Zargar, and Iraj Kazeminezhad. "Colorimetric disposable paper coated with ZnO@ ZnS core–shell nanoparticles for detection of copper ions in aqueous solutions." *ACS Applied Materials & Interfaces* 6, no. 20 (2014): 17694–17701. doi:10.1021/am505480y.

55. Zhi, Lihua, Wei Zuo, Fengjuan Chen, and Baodui Wang. "3D MoS_2 composition aerogels as chemosensors and adsorbents for colorimetric detection and high-capacity adsorption of Hg^{2+}." *ACS Sustainable Chemistry & Engineering* 4, no. 6 (2016): 3398–3408. doi:10.1021/acssuschemeng.6b00409.

56. Dang, Qin-Qin, Hong-Jing Wan, and Xian-Ming Zhang. "Carbazolic porous framework with tetrahedral core for gas uptake and tandem detection of iodide and mercury." *ACS Applied Materials & Interfaces* 9, no. 25 (2017): 21438–21446. doi:10.1021/acsami.7b04201.

57. He, Dinggeng, Xiaoxiao He, Kemin Wang, Yingxiang Zhao, and Zhen Zou. "Regenerable multifunctional mesoporous silica nanocomposites for simultaneous detection and removal of mercury (II)." *Langmuir* 29, no. 19 (2013): 5896–5904. doi:10.1021/la400415h.

58. Peng, Xiaohong, Yujiao Wang, Xiaoliang Tang, and Weisheng Liu. "Functionalized magnetic core–shell Fe_3O_4@SiO_2 nanoparticles as selectivity-enhanced chemosensor for Hg (II)." *Dyes and Pigments* 91, no. 1 (2011): 26–32. doi:10.1016/j.dyepig.2011.01.012.

59. Huang, Haowen, Caiting Qu, Xuanyong Liu, Shaowen Huang, Zhongjian Xu, Bo Liao, Yonglong Zeng, and Paul K. Chu. "Preparation of controllable core–shell gold nanoparticles and its application in detection of silver ions." *ACS Applied Materials & Interfaces* 3, no. 2 (2011): 183–190. doi:10.1021/am101034h.

60. Wang, Hengguo, Lei Sun, Yapeng Li, Xiaoliang Fei, Mingda Sun, Chaoqun Zhang, Yaoxian Li, and Qingbiao Yang. "Layer-by-layer assembled Fe_3O_4@ C@ CdTe core/shell microspheres as separable luminescent probe for sensitive sensing of Cu^{2+} ions." *Langmuir* 27, no. 18 (2011): 11609–11615. doi:10.1021/la202295b.

61. Kim, Changwoo, Seung Soo Lee, Benjamin J. Reinhart, Minjung Cho, Brandon J. Lafferty, Wenlu Li, and John D. Fortner. "Surface-optimized core–shell nanocomposites (Fe_3O_4@ $Mn_xFe_yO_4$) for ultra-high uranium sorption and low-field separation in water." *Environmental Science: Nano* 5, no. 10 (2018): 2252–2256. doi:10.1039/C8EN00826D.

62. Du, Yuanxin, Renyong Liu, Bianhua Liu, Suhua Wang, Ming-Yong Han, and Zhongping Zhang. "Surface-enhanced Raman scattering chip for femtomolar detection of mercuric ion (II) by ligand exchange." *Analytical Chemistry* 85, no. 6 (2013): 3160–3165. doi:10.1021/ac303358w.

63. Fu, Qiangqiang, Ze Wu, Dan Du, Chengzhou Zhu, Yuehe Lin, and Yong Tang. "Versatile barometer biosensor based on Au@ Pt core/shell nanoparticle probe." *ACS Sensors* 2, no. 6 (2017): 789–795. doi:10.1021/acssensors.7b00156.

8 Quantum Dots as Plasmonic Nanosensors

8.1 INTRODUCTION

The fabrication of sensitive, selective, high-throughput, practical, and workable analysis must recognize inorganic pollutants. Various techniques, such as inductively coupled plasma mass spectrometry (ICPMS), AAS, and atomic emission spectrometry (AES), have high accuracies and sensitivity [1] toward previous metal-specific-based methods. These methods require sophisticated operators, wide-ranging methods for sample pretreatment, bulky instrumentation, and limit their applications in in situ analysis. Various innovative detection platforms based on comparatively low-cost and straightforward operative paths are substitutes, such as SPR detections, electrochemical methods, fluorescence, CL, quartz crystal microbalance methods, etc. [2–4]. Of these, fluorescence techniques benefit relative simplicity, high accuracy, and high sensitivity. For example, performing fluorescent sensors, for example, reliability, sensitivity, reproducibility, and dynamic range, depends on choosing the suitable fluorophore. Various key factors such as fluorescence QY, fluorescence lifetime, the shape and width of its emission and absorption bands, molar absorption coefficients, emission and transmission band maxima, and Stokes's shift will affect the fluorophore performance for the detection of HMs. Below the optimum excitation wavelength, an efficient fluorophore should be bright with a high fluorescence QY and high molar absorption coefficient. Different fluorophores can be developed combined or separately as HMs sensing platforms such as metal–ligand complexes, small organic dyes, metal nanoclusters (NCs) and semiconductors [5], MOFs [6,7], 0D carbon dots (CDs) [8], and silica NPs [9]. Small molecular organic dyes, such as fluorescein, rhodamine, coumarin, and cyanine, are the utmost in broadly used and traditional fluorophore. However, their drawbacks comprise the emission or excitation wavelengths, small around 25 nm stokes shifts simply pretentious thru the self-quenching effect, and environmental variations limiting their applications. CDs are a substitute and excellent for fluorescent sensors toward HMs, in contrast with other fluorophores.

The fluorescence as their instinct properties and carbon-based nanosized materials with <10 nm of characteristic sizes is known as CDs [10,11]. According to their elementary structures, they can be categorized into three types. The carbon NPs of spherical shape with nanocrystalline or amorphous cores are known as CQDs and carbon nanodots (CNs). The CDs holding graphene layers with chemical groups make the GQDs, and polymer dots (PDs) are the CDs with polymer cross-linking or grafting on their edges. With CL or fluorescence detection, CDs

DOI: 10.1201/9781003128281-8

are excellent electron donors and electron acceptors usually used to detect metal ions. Luminescence is interrelated to EMR generated in the CL system during a chemical reaction, and no photoexcitation is required [12]. CDs can be utilized as emitting species, oxidants, even catalysis for chemical reaction energy. These work as energy acceptors [13–16]. Photoexcitation is essential for good signals with fluorescence detection to produce CDs and photoluminescence (PL) emissions. The overall merits can be characterized as follows as sensors for HMs as CDs. Foremost, CDs own optical fluorescent properties with low photobleaching, increased photostability, and therefore for fluorescence detection of HMs, these are appropriate as fluorophores.

Most CDs in aqueous media are hydrophilic with good solubilities to empower their promising use as a sensor for HMs in the medical, biotic, and environmental fields [17]. For HMs sensing in bacterium, cells, animals, and plants, most CDs are used as fluorescent probes with great potential applications and display low toxicity with their synthesized biocompatible functional groups and their carbon core on the surface. Recently, for sensing of HM ions, their novel applications in the testing of food safety and ecological species [18], chemical logic gates [19], animal and cell imaging [20,21], document counterspy [22], fingerprint security, and anti-counterfeiting ink [23,24], various significant developments on CDs have been made as fluorescence sensors. This chapter focuses on the current developments on CDs as fluorescence sensing platforms comprising CQDs, CNs, carbon nitride dots (CNDs), C-dots, PDs, and GQDs to detect HMs. The detection of heavy metals, PL principles, and the methods for binding among HMs and CDs has briefly presented the techniques to purify and synthesize CDs. The development on detection of diverse metal species, single HM ions, and multi-HMs are disapprovingly retrieved from the signal detection mode. The applications in the bacterial, cell, animal, and plant bioimaging of HMs, the progress of sensor platforms of solid-state presented as CDs. This chapter focuses on different quantum dot-based detection probes for detecting toxic heavy metals.

8.2 APPLICATIONS OF QUANTUM DOTS FOR THE SENSING OF HEAVY METAL IONS

For heavy metal ions detection, an effective and simple method has been described for designing ratiometric fluorescent nanosensors. Thru coating of CDs on silica NPs doped with Rhodamine B (RhB) surface, a CD-based dual emission nanosensor was synthesized for the sensing of Cu^{2+} ions. On the surface of CDs, the residual methoxysilane groups and ethylenediamine groups can serve as the silylation reaction groups and the detection sites for Cu^{2+} ions. N-(β-aminoethyl)-γ-aminopropyl methyldimethoxysilane (AEAPMS) is the key raw material utilized to synthesize fluorescent CDs. Below a single wavelength for excitation, the attained nanosensor displayed distinctive emissions of fluorescence of CDs (blue) and RhB (red). Due to the quenching of CDs fluorescence, the response of ratiometric fluorescence occurs of the dual emission silica NPs and the binding with Cu^{2+} ions. In the presence of amino acids, heavy metals, vitamin C, and proteins, etc., the developed ratiometric nanosensor showed virtuous selectivity for Cu^{2+} ions. The LOD was found to be 35.2 nM in the range of $0–3 \times 10^{-6}$ M of the Cu^{2+} ions, with a linear decrease in the ratio of F467/F585. The developed nanosensor was fruitfully applied for sensing in real and tap water and ratiometric fluorescence imaging of Cu^{2+} in cells [25].

Owing to its potential hazards toward environmental stability and human safety, an efficient and sensitive platform has been designed for analysis of Hg^{2+} ion is of great significance. For the fluorescent sensing of Hg^{2+} ions, the MoO_{3-x} QDs have been developed. The probe functions on the synergistic effect of Hg^{2+} on cobalt porphyrin formation and MoO_{3-x} fluorescent QDs. Please reform the sentence as" The results of synergistic interaction in rapid and distinct development of the optical properties of MoO_{3-x} QDs and porphyrins facilitates the intracellular Hg^{2+} imaging and their kinetic sensing [26].

Through the synergistic effect of trace Hg^{2+} ions and GQDs doped with nitrogen (NGQDs), to magically advance the complexation reaction rate among Mn^{2+} ion and porphyrin, a simple method has been proposed. For attacking small divalent metal ions, a mechanism is anticipated to be carried by NGQDs from the back consistent with a substitution reaction in which thru relatively larger Hg^{2+} ions, the deformed porphyrin nucleus is favorable. In the meantime, the metalloporphyrin formation is escorted thru the fluorescence quenching of porphyrins and absorption red shift. Concurrently, owing to the inner filter consequence among NGQDs and porphyrins, the fluorescence of NGQDs is progressively improved. Therefore, the colorimetric methods and ratiometric fluorescence have been projected for the sensing of trace Hg^{2+} ions. These methods have potential application in biological and complex environmental conditions and are based on the discrete fluorescence/absorption spectral changes [27].

Similarly, carbon nitride materials have developed exceedingly discovered nanomaterials based on carbon because of their semiconductor properties. Therefore, a facile one-pot synthesis from ethylenediaminetetraacetic acid disodium salt and thiourea, the quantum dots of carbon nitride doped with sulfur and oxygen (using a thermal method) SCNQDs) has been synthesized. Various analytical tools have been used to characterize the developed SCNQDs, that is, UV–vis, TEM, FTIR, XRD, and powder XPS. The sensing ability of SCNQDs has been established for Hg^{2+} ion sensing application in solid. Specifically, SCNQDs are loaded onto filter paper and the solution phase. In the 10 nM to 1 μM concentration range, the sensitivity of SCNQDs showed a linear relationship for Hg^{2+} ion. The lowest limit detection of the developed probe was found to be 0.01 nM. Likewise, the filter paper loaded with SCNQDs has been checked for sensing capability toward Hg^{2+} ions in Millipore water as Hg^{2+} in solutions and in spiked tap water. To detect room temperature phosphorescence (RTP) in SCNQDs, a poly(vinyl alcohol) (PVA) used SCNQDs composite has been developed [28].

For an accurate and on-site effective detection of inorganic pollutants, a solid, robust detection platform is still challenging. An agarose hydrogel film based on chitosan-based CDs has been synthesized as a solid hybrid sensing platform for heavy metal ions. The fabricated detection probe shows the electrostatic interaction between the NH_4^+ group and OH^- groups of CDs and agaroses, correspondingly. The solution of heavy metals, such as Cu^{2+}, Cr^{6+}, Pb^{2+}, Mn^{2+}, Fe^{3+}, shows a color alteration of hydrogel film strip on simply dipping it. The developed probe showed the LOD up to 1 pM, 0.5 μM, and 0.5 nM for Cu^{2+}, Cr^{6+}, Pb^{2+}, Mn^{2+}, and Fe^{3+}. The hydrogel film shows efficient applicability for eliminating this quintet of heavy metals as a filtration membrane. The synthesized hybrid hydrogel probe is most suitable for cheap, portable, and on-site effective optical colorimetric detection probes for heavy metals [29].

Similarly, for the sensing of glutathione (GSH) with high sensitivity besides Hg^{2+} ions, a specific ratiometric fluorescent probe have been developed using a reference, that is, RhB and a sensing signal, that is, CDs. Attained nanosensor shows characteristic fluorescence emission of CDs and RhB under a single wavelength required for excitation. According to a one-pot pyrolysis process, the operative CDs were fabricated in the laboratory using histidine and sodium citrate. Because of surface functional groups on CDs and transferring electrons among Hg^{2+} ions, the CDs' fluorescence was reduced in the presence of the Hg^{2+} ion. Then, because of their stronger affinity with Hg^{2+}, the fluorescence of CDs–Hg^{2+} system efficiently exclude several background interferences and improved progressively with the addition of GSH. To oxidative stress model investigation, this method has been fruitfully utilized. The developed ratiometric detection method shows the enormous potential for detecting GSH and Hg^{2+} ions in biological detection and environmental monitoring, which is reliable, facile, selective, and sensitive [30].

In analytical detection and biological imaging, semiconducting PDs hold great promise and are proved as an innovative fluorescent ultrabright sensor. In this instance, a Pdots-based visual sensor is used to recognize Pb^{2+} ions in solution. The development done thru the incorporation of NIR dyes and poly[(9,9-dioctylfluorene)-co2,1,3-benzothiadiazole-co-4,7-di(thiophen-2-yl)-2,1,3-benzothiadiazole] (PFBT-DBT) into the polymer matrix. After that, it was further capped with the pre-functionalized polydiacetylenes (PDAs) with 15-crown-5 moieties. Because of the development of the 15-crown-5-Pb^{2+}-carboxylate sandwich complex in the 2:1 ratio on the surface of Pdot, these developed Pdots are highly selective for Pb^{2+} ions. The change in color from blue to red of the PDA occurs after chelation with Pb^{2+} ions, and the PDA conjugation system was strained and perturbed.

The test strips are made of Pdot-poly(vinyl alcohol) film used for in situ on-site Pb^{2+} ion sensing in real aqueous media [31]. Similarly, the CdS (COF-CdS) QDs functionalized with carboxyl were developed to sense Co^{2+} ions colorimetrically in an aqueous solution. After the interaction of the developed probe with Co^{2+} ions, the color changes from colorless to yellowish-brown within 5 min, which naked eyes can notice. After the addition of Co^{2+} ions, the COF-CdS QDs got aggregated, which is further confirmed by enhancing the UV–vis absorption spectra at 360 nm. Under the optimized conditions, the developed method showed LOD of 0.23 μg mL^{-1} for Co^{2+} ions with outstanding sensitivity and selectivity.

In contrast to the Co^{2+} ion concentration having a linear range of 0.5–14 μg mL^{-1}, the calibration plot of $(A - A_0)$ shows a correlation coefficient at

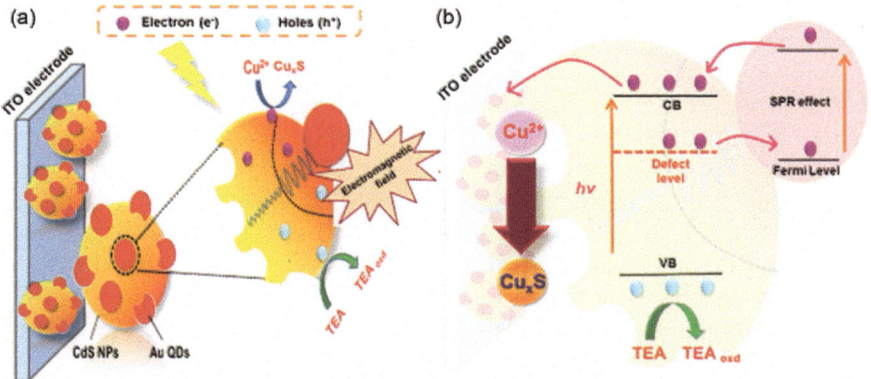

FIGURE 8.1 Diagrammatic illustration of (a) AuQDs, the electromagnetic field interaction with CdS NPs (b) possible mechanism for Cu^{2+} ion sensing. (Reprinted with permission from Ref. [38]. Copyright 2016 American Chemical Society.)

360 nm of 0.9996. The technique's reliability and accuracy were further determined through the standard addition method through recovery studies in the range of 99.63%–102.46% with percent recoveries. In aqueous environmental samples, this technique offers a simple, fast, and cost-efficient solution for sensing Co^{2+} ions in the existence of a complex matrix [32]. Figure 8.1 shows the schematic representation of the mechanism for Cu^{2+} ion sensing.

A phosphorescent Mn-doped ZnS QDs detection probe has been fabricated to detect Cr^{3+} ions by the capping of denatured bovine serum albumin (dBSA). Due to the electron transfer to Cr^{3+} from photoexcited Mn-doped ZnS QDs, the phosphorescent quenching mechanism occurs based on phosphorescence decay and calculations of the relative energies of Cr^{3+} and QDs. In the range of 10–300 nM, virtuous linear quenching Stern–Volmer equation was attained for Cr^{3+} ions under the optimal conditions. The LOD was found to be 3 nM of the developed phosphorescence probe from the sample matrix due to the effective scattering and removal of background fluorescence. In the aqueous samples to determine Cr^{3+} ions, the developed probe showed analytical potential ranging from 95% to 106% with spike–recoveries [33]. In another study, for easy elimination and sensing of heavy metals based on cation exchange reaction, such as Pb^{2+}, Ag^+, and Hg^{2+} in water, ZnS quantum dots stabilized with biopolymer QDs were developed. ZnS QDs stabilized with chitosan were synthesized in the existence of corresponding ions, which further changed to PbS, Ag_2S, and HgS Q-dots in aqueous systems. Because of the visible bright yellow color formation, the converted Q-dots exhibited a distinguishing development of color with Hg^{2+} ions, while in additional metals, the brown coloration was detected. The alteration in the product solubility and the reactant QDs, thru the cation exchange mechanism. Based on the parameters of a bulk crystal lattice, the cation exchanged Q-dots showed increased volume and conserved the morphology of the reactant Q-dots. Subsequently, by the AAS results, it is confirmed that the developed chitosan film impregnated ZnS Q-dot probe exhibited excellent removal of hazardous metals from water samples [34].

Similarly, for bioimaging and target analysis, a CD fluorescent detection probe has been fruitfully developed, conjugating through amide bonds and thiosemicarbazide. The size distribution of the functional CDs is relatively about 1.6–4.0 nm. In the presence of competing metal ions, the CDs show exceedingly precise detection efficiency, and the rate of maximum quenching ranges to 68.7% toward copper ions Cu^{2+}. The probe showed the LOD at 3.47 nM, in the concentrations range 0–0.4 μM of Cu^{2+} ions with an excellent linear correlation among F_0/F. The nanosensor establishes its bio labeling potential in vivo due to its excellent water solubility, strong fluorescence intensity, good biocompatibility, and low toxicity, which can be highly important in biomedical detection and bioanalysis soon [35].

Similarly, due to the utmost intense toxicity of Hg^{2+} at trace level to living things and human health, a water-soluble, cheap, and rapid metal sensor has been developed. For the ultra-selective and sensitive sensing of Hg^{2+} ions ranging from 0 to 20 nM with the lowest LOD up to 0.1 nM, a metal ion sensor was used made of mercaptopropionic acid (MPA) coated ZnSe/ZnS colloidal NPs doped with Mn. The MPA was utilized as an exceedingly inimitable acceptor for Hg^{2+} ions due to the strong interaction between mercury ions and thiol(s). Because of variations of surface chemical properties, colloidal NPs quickly agglomerated, which results in severe quenching of fluorescent intensity in the existence of mercury ions. The QD-based metal ions sensor can be utilized to examine real water samples with high sensitivity and selectivity, and satisfactory accuracy [36].

In addition, for rapid and simultaneous detection of multiple analytes, the development of cost-effective and versatile detection methods is exceedingly required for home healthcare, food safety inspection, and environmental observation. In this instance, CDs dependent paper-based microarrays based on a fluorescence turn-off sensing system and a custom-designed APP (SBR-App) installed stand-alone smartphone-based portable reader (SBR) have been developed to detect the analyte. The designed probe simultaneously detects heavy metals Cu^{2+}, Pb^{2+}, and Hg^{2+} ions in the aqueous systems. Using this method, except for a smartphone, there is no prerequisite for expensive equipment and complex sample pretreatment. For an easy-to-handle and biodegradable detection point of care, this versatile and cost-efficient smartphone-based detection method featured simplicity and reliability and preferably matched in resource-constrained environments [37].

In another study, the abnormal ingestion of Cu^{2+} ion displays acute toxicity and has important adverse properties in living organisms. In contrast to other conventional methods, owing to its low cost, high selectivity, sensitivity, and precise selection, the photoelectrochemical (PEC) method has fascinated ample consideration for detecting Cu^{2+} as an ion sensor. For the synthesis of cadmium sulfide NPs (CdS NPs), in situ and stepwise hydrothermal chemical methods have been used. The developed NPs used to decorate gold quantum dots (AuQDs), accompanied by notable PEC performance. For photoexcited electrons, the allotted cooking time is 268 fs for CdS and 243 fs for Au, shown by the ground state recovery femtosecond transient absorption dynamics. With no contribution from

Au, the CdS NPs-AuQDs transient spectrum was conquered through a CdS signal. From CdS to Au, a fast transfer of the photoexcited electrons is shown by the absence of rapid accumulation dynamics before cooling down in CdS-Au. The detection of Cu^{2+} ions occurs in a linear range of 0.5–120 nM with 6.73 nM of detection limit using developed CdS NPs-AuQDs-2 photoelectrode. The CdS NPs-AuQDs-2 were further tested for selectivity for Cu^{2+} ions in mixed metal ion solutions, which showed promise as a photoactive material for PEC-based environmental analysis and monitoring in the lake and tap water [38].

Similarly, for biosensing and bioimaging applications, the QD emitters are the potential sensors due to their chemical stability, bio functionality, and biocompatibility. Achieving these criteria of effective emission of NIR using nontoxic QDs remnants was exceedingly challenging. Therefore, water-soluble $AgInS_2$/ ZnS core/shell (AIS/ ZnS) QDs functionalized with DNA emitting NIR have been developed to overcome these issues. At 700 nm, 55% of an extraordinary photoluminescence quantum yield (PLQY) and 900 ns of long PL decay time, depending on a two-step synthesis of hot injection. The developed aqueous route corresponds to exceedingly luminescent chalcopyrite-type AIS/ZnS core/shell QDs. It yielded a diverse performance as size distribution, crystal structure, and stoichiometry related the slow and fast precursors hot injection to AIS core QD development. Both core QDs showed PLQYs of 8% and 36%, with PL peak positions of 760 nm (slow injection) and 710 (fast). The successive and slow integration of the S and Zn precursors endorsed the ZnS shell formation with the thickness of a three-monolayer. The upsurge from 3.0 to 4.8 nm of the average size of QD occurs due to the succeeding shell growth step on the sturdier emitting cores. Bioconjugation of the AIS/ZnS QDs during the ZnS shell growth was achieved with DNA changed with hexylthiol, after a 5–6 DNA single strands grafting level per QD. The agarose gel electrophoresis and UV–vis spectroscopy were used to attest to the successful chemical conjugation of DNA. Significantly, the stability of the AIS/ZnS-DNA QD conjugates and their fruitful coupling corroborated using complementary DNA strands by SPR imaging experiments, along with the preservation of anchored DNA biological activity. For their usage in biomedical applications, a high potential is provided by the strong NIR emission and biocompatibility of the developed QDs of AIS/ ZnS-DNA [39].

Similarly, carbon dots ($CuDTC_2$-CDs) functionalized with bis-(dithiocarbamato)copper(II) have been fabricated to recognize Hg^{2+} ions selectively. Using condensation of carbon disulfide on the nitrogen atoms existing on amine groups surface, the $CuDTC_2$ complex first conjugated with the prepared amine-coated CDs to form the $CuDTC_2$-complexing CDs. This complex binds with resulting dithiocarbamate groups (DTC) together with the copper(II) followed by ammonium N-(dithicarbaxy) sarcosine (DTCS) coordination. At the surface, the $CuDTC_2$ complex, a combination of energy and electron transfer methods, sturdily reduced the fluorescence of the CDs of bright blue. The $CuDTC_2$ complex instantly dislocates the Cu^{2+} ions after adding the Hg^{2+} ions, and it could instantly switch on the $CuDTC_2$-CDs fluorescence. This process shuts the pathway of energy transmission down with the LOD of 4 ppb for Hg^{2+} ions.

CuDTC$_2$-CDs probe ink resulted in a paper-based sensor through the printing process using a commercial inkjet printer on a piece of cellulose acetate paper. The Hg^{2+} ions can be detected by the developed "turn on" fluorescence paper by the naked eye [40]. In addition, for the Hg^{2+} ions recognition in a water sample, long lifetime fluorescence quantum dots soluble in water and AuNPs embedded in a time-gated fluorescence resonance energy transfer (TGFRET) detection probe has been developed. Excluding four intentionally fabricated T–T mismatches, the two complementary ssDNA functionalized AuNPs together with fluorescence quantum dots soluble in water have been fabricated. The AuNPs and QDs acted as the acceptor and donor for the energy transfer process, respectively. Owing to the T–Hg^{2+}–T complex formation, DNA hybridization will occur in the existence of Hg^{2+} ions in the aqueous solution. The fluorescence intensity of quantum dots decreases apparently because of the energy transmission from QDs to AuNPs by the proximity of Au NPs and quantum dots. As the Hg^{2+} ions concentration is increased, the fluorescence intensity proportionally decreases. In tap water, the developed probes showed 0.87 nM of detection limit, whereas, in buffer samples, it is 0.49 nM under the optimum conditions for Hg^{2+} ions ranging from 1×10^{-8} to 1×10^{-9} M. The Hg^{2+} ion sensing was done using this sensor with good results in Hg^{2+} ions spiked lake, river, and tap water [41].

Similarly, oyster mushroom (*Pleurotus* species) functionalized fluorescent blue/green CQDs have been developed thru a one-step simplistic hydrothermal carbonization method. The synthesized C-dots were utilized as a colorimetric sensor for Pb^{2+} ion sensing with the LOQ and LOD of 177.69 and 58.63 μM, correspondingly. It is also utilized as a fluorescent probe to recognize DNA by the electrostatic interactive interaction among CDs and ctDNA. The fabricated CQDs showed effective antibacterial activity for three bacterial strains: *Pseudomonas aeruginosa*, *Klebsiella pneumoniae*, *Staphylococcus aureus*, and the CDs show anticancer activity against MDA-MB-231 breast cancer cells [42].

Similarly, owing to their excellent PL properties, good optical stability, and less cytotoxicity, the GQDs have attracted considerable attention. Therefore, exceedingly fluorescent nitrogen-doped graphene quantum dots (NGQDs) have been fabricated to take triethanolamine and sodium citrate in microwave heating as starting substances. The developed N-GQDs exhibited pH-sensitive properties and excellent stability, outstanding consistent distribution with 5.6 nm of average diameter, and displayed a significant QY of 8% of ensuring bright blue fluorescence. The N-GQDs selectively detect Fe^{3+} ions owing to the static quenching mechanism in aqueous media with 9.7 nM of LOD [43].

To recognize Cu^{2+} ions with ultrahigh selectivity and improved sensitivity, a cost-effective and simple QDs- based detection technique has been fabricated in the presence of thiosulfate. Therefore, a CdSe/ZnS QDs fluorescent probe changed with CTAB is used. However, the sensitivity could be attained in a few nanomolar, whereas silver and mercury ions display strong interference in the thiosulfate absence with Cu^{2+} ions. In the existence of thiosulfate, the detection limit is 0.15 nM for Cu^{2+} ions in distilled water and 0.14 nM in a tap water sample.

The developed probe showed a virtuous potential for Cu^{2+} ions with rapidity, simplicity, excellent selectivity, and ultrahigh sensitivity [44].

In another study, luminescent porous silicon (LuPSi) with a photoluminescent 2D PL sensing method having abundant surface chemistry and silicon quantum dots of wide-size distribution. The reaction or interaction between LuPSi and analytes can cause dynamic, static, deposition-induced, and oxidation-induced quenching because of the intrinsic nature of LuPSi. The 2D PL sensing method can favor diverse analytes by monitoring both the peak shift of LuPSi and PL intensity change. The 2D PL method is based on suggestive ease of the detection and sensing elements, in contrast to the existing array-based methods [45]. Similarly, various quantum dots have been incorporated in Table 8.1 to detect toxic metals in water systems using different methods.

8.3 CONCLUSION AND FUTURE PERSPECTIVES

Owing to the promising properties of QDs over metal oxide/noble metal or fluorescent dyes nanomaterials, QD-based fluorescence sensors are outstanding techniques for imaging or sensing heavy metal ions. Various QD-based fluorescence sensors have been reported that a tremendous capacity to sense heavy metal ions is based explicitly on ratiometric and "turn-on" measurement. To increase the detection output, sensing of heavy metal ions comprising high toxic species, development of solid/gel sensor platforms such as sensor arrays, paper-based devices, and polymer films or hydrogel, as an intelligent visual, cost-effective, in vivo, or in vitro, and on-site analysis for bioimaging of heavy metal ions. Conclusively, to detect heavy metal ions, a great advancement toward the fabrication of QD-based sensing systems. However, in the future, many challenges need to be overcome using emerging trends given subsequently:

- The manageable fabrication of QDs with desirable size and uniformity is still lacking after the intricacy of the indistinct origin of their PL emission and composition.
- This can significantly affect QDs and optical feature activity as heavy metal detection probes and hinder the final application in the market and commercialization.
- Correspondingly, after carbonization and polymerization, the present method cannot recognize the precise molecular structure of QDs and describe the HMs-QDs coordinated complexes. Additional works are essential to advance the fabrication techniques to attain size-controlled QDs and to characterize the developed QDs adequately.
- The difficulty in purifications with low product yield is also a problem, preventing the wide-ranging manufacture of QDs as sensors for detecting heavy metal ions. More efforts must be made to discover easy-to-operate, reproducible, and fast purification techniques and upsurge product yield.

TABLE 8.1

Detection of Heavy Metal Ions Using Different Quantum Dot-Based Materials

Detection Probe	Analyte	Method	LOD	Size	Quantum Yield (%)	Linear Range	References
Dopamine functionalized GQDs (DA-GQDs)	Fe^{3+}	Fluorescence	7.6 nM	4.5±0.6	10.2	20 nM – 2 µM	[46]
Cadmium sulfide (CdS) quantum dots	Ag^+	-	-	-	-	-	[47]
Carbon quantum dots (CQDs)	Cu^{2+}	Electrochemical	1×10^{-14} M	-	-	-	[48]
Coal-derived graphene quantum dots (C-GQDs)	Cu^{2+}	Fluorescence	-	3.2±1.0 nm	-	-	[49]
Poly(ethylene glycol) passivated graphene quantum dots (PEG-GQDs)	Fe^{3+}	Fluorescence	5.77 µM	2.5 nm	21.32	-	[50]
Cadmium telluride (CdTe) quantum dots (QDs)	Hg^{2+}	Fluorescence	2.63 nM L^{-1}	10 nm	-	10–60 nM	[51]
16-Mercaptohexadecanoic acid (16-MHA) capped CdSe quantum dots (QDs)	Cu^{2+}	Fluorescence	5 nM	-	-	5 nM to 100 µM	[52]
$CH_3NH_3PbBr_3$@MOF-5 composites	-	Fluorescence	-	-	-	-	[53]
Eu^{3+}-doped lead-free $Cs_3Bi_2Br_9$ perovskite quantum dots (PeQDs)	Cu^{2+}	Fluorescence	10 nM	-	18 to ~42.4	5 nM to 3 µM	[54]
Aminophosphine-based InP QDs	Zn^{2+} and Cd^{2+}	Fluorescence	-	-	15 times together with 55 nm red shift	-	[55]
Branched poly(ethylenimine) (BPEI)-functionalized carbon quantum dots (CQDs)	Cu^{2+}	Fluorescence	6 nM	-	-	10–1,100 nM	[56]
Graphene quantum dots (GQDs)	Fe^{3+}	Fluorescence	0.23 µM	1–5 nm	-	-	[57]
Azamacrocycle derivatization of CdSe/ZnS core/shell quantum dot	Zn^{2+}	Fluorescence	2.4 µM	-	-	5–500 µM	[58]
ZnS:Ce QDs	Hg^{2+}	Fluorescence	0.82 µM L^{-1}	-	-	10–100 µM	[59]

- QD-based test paper arrays or strips have outstanding potential for monitoring HMs and route quality control as practical kits for HMs sensing purposes, especially in ecological submissions.
- Developments in the manufacture of dosage-sensitive, stable, and simple paper strips with high performance like pH test paper to detect heavy metal ions are expected in the future. Therefore, to manufacture inexpensive and portable devices for on-site detection, more attention should be paid.
- In single toxic metal detection, the overwhelming reports display excessive performance.
- Though leaving an enormous open space for future developments to detect diverse species of particular heavy metals and multi-heavy metals, the research advancement on QDs is still in its infancy.
- QDs functionalized with enzymes and small oligonucleotide ligands (aptamer) to upsurge the selectivity for in vivo sensing of heavy metal, with strong and specific affinities to target heavy metal will be auspicious methods.
- So far, with a comparatively simple matrix-like water, utmost the fluorescent QD-based sensors are applied to the real sample for heavy metal analysis.
- The matrices are much more complex, with many organic compounds and ions for most biological and environmental samples, which can cause deterioration or impairment in the performance of fluorescent sensors. To detect HMs in complex matrices, the design of stable and reliable QD-based sensors is still a significant challenge.

Concluding, in bioimaging and sensing applications, to overwhelm the challenges mentioned above, QDs have established prodigious capability to sense heavy metals and their species. A long-term effort is still required in this field.

REFERENCES

1. Suvarapu, Lakshmi Narayana, and Sung-Ok Baek. "Recent developments in the speciation and determination of mercury using various analytical techniques." *Journal of Analytical Methods in Chemistry* 2015 (2015). doi:10.1155/2015/372459.
2. Yang, Shen, Hu Jiwen, Liu Tingting, Gao Hongwen, and Hu Zhangjun. "Colorimetric and fluorogenic chemosensors for mercury ion based on nanomaterials." *Progress in Chemistry* 31, no. 4 (2019): 536. doi:10.7536/PC180933.
3. Fen, Yap Wing, and W. Mahmood Mat Yunus. "Surface plasmon resonance spectroscopy as an alternative for sensing heavy metal ions: a review." *Sensor Review* (2013). doi:10.1108/SR-01-2012-604.
4. Khanmohammadi, Akbar, Arash Jalili Ghazizadeh, Pegah Hashemi, Abbas Afkhami, Fabiana Arduini, and Hasan Bagheri. "An overview to electrochemical biosensors and sensors for the detection of environmental contaminants." *Journal of the Iranian Chemical Society* (2020): 1–19. doi:10.1007/s13738-020-01940-z.

5. Oliveira, Elisabete, Cristina Núñez, Hugo Miguel Santos, Javier Fernández-Lodeiro, Adrián Fernández-Lodeiro, José Luis Capelo, and Carlos Lodeiro. "Revisiting the use of gold and silver functionalized nanoparticles as colorimetric and fluorometric chemosensors for metal ions." *Sensors and Actuators B: Chemical* 212 (2015): 297–328. doi:10.1016/j.snb.2015.02.026.

6. Samanta, Partha, Sumanta Let, Writakshi Mandal, Subhajit Dutta, and Sujit K. Ghosh. "Luminescent metal–organic frameworks (LMOFs) as potential probes for the recognition of cationic water pollutants." *Inorganic Chemistry Frontiers* 7, no. 9 (2020): 1801–1821. doi:10.1039/D0QI00167H.

7. Rasheed, Tahir, Muhammad Bilal, Faran Nabeel, Hafiz MN Iqbal, Chuanlong Li, and Yongfeng Zhou. "Fluorescent sensor-based models for the detection of environmentally-related toxic heavy metals." *Science of the Total Environment* 615 (2018): 476–485. doi:10.1016/j.scitotenv.2017.09.126.

8. Ahmad, Rafiq, Nirmalya Tripathy, Ajit Khosla, Marya Khan, Prabhash Mishra, Waquar Akhter Ansari, Mansoor Ali Syed, and Yoon-Bong Hahn. "Recent advances in nanostructured graphitic carbon nitride as a sensing material for heavy metal ions." *Journal of the Electrochemical Society* 167, no. 3 (2019): 037519. doi:10.1149/2.0192003JES.

9. Kumar, Pawan, Ki-Hyun Kim, Vasudha Bansal, Theodore Lazarides, and Naresh Kumar. "Progress in the sensing techniques for heavy metal ions using nanomaterials." *Journal of Industrial and Engineering Chemistry* 54 (2017): 30–43. doi:10.1016/j.jiec.2017.06.010.

10. Xu, Xiaoyou, Robert Ray, Yunlong Gu, Harry J. Ploehn, Latha Gearheart, Kyle Raker, and Walter A. Scrivens. "Electrophoretic analysis and purification of fluorescent single-walled carbon nanotube fragments." *Journal of the American Chemical Society* 126, no. 40 (2004): 12736–12737. doi:10.1021/ja040082h.

11. Duran, Nelson, Mateus B. Simoes, Ana de Moraes, Wagner J. Favaro, and Amedea B. Seabra. "Nanobiotechnology of carbon dots: A review." *Journal of Biomedical Nanotechnology* 12, no. 7 (2016): 1323–1347. doi:10.1166/jbn.2016.2225.

12. Wei, Jian-Yong, Qing Lou, Jin-Hao Zang, Zhi-Yu Liu, Yang-Li Ye, Cheng-Long Shen, Wen-Bo Zhao, Lin Dong, and Chong-Xin Shan. "Scalable synthesis of green fluorescent carbon dot powders with unprecedented efficiency." *Advanced Optical Materials* 8, no. 7 (2020): 1901938. doi:10.1002/adom.201901938.

13. Shen, Cheng-Long, Qing Lou, Kai-Kai Liu, Lin Dong, and Chong-Xin Shan. "Chemiluminescent carbon dots: Synthesis, properties, and applications." *Nano Today* 35 (2020): 100954. doi:10.1016/j.nantod.2020.100954.

14. Shen, Cheng-Long, Guang-Song Zheng, Meng-Yuan Wu, Jian-Yong Wei, Qing Lou, Yang-Li Ye, Zhi-Yi Liu, Jin-Hao Zang, Lin Dong, and Chong-Xin Shan. "Chemiluminescent carbon nanodots as sensors for hydrogen peroxide and glucose." *Nanophotonics* 1, no. ahead-of-print (2020). doi:10.1515/nanoph-2020-0233.

15. Shen, Cheng-Long, Qing Lou, Jin-Hao Zang, Kai-Kai Liu, Song-Nan Qu, Lin Dong, and Chong-Xin Shan. "Near-infrared chemiluminescent carbon nanodots and their application in reactive oxygen species bioimaging." *Advanced Science* 7, no. 8 (2020): 1903525. doi:10.1002/advs.201903525.

16. Shen, Cheng-Long, Qing Lou, Chao-Fan Lv, Jin-Hao Zang, Song-Nan Qu, Lin Dong, and Chong-Xin Shan. "Bright and multicolor chemiluminescent carbon nanodots for advanced information encryption." *Advanced Science* 6, no. 11 (2019): 1802331. doi:10.1002/advs.201802331.

17. Boakye-Yiadom, Kofi Oti, Samuel Kesse, Yaw Opoku-Damoah, Mensura Sied Filli, Md Aquib, Mily Maviah Bazezy Joelle, Muhammad Asim Farooq et al. "Carbon dots: Applications in bioimaging and theragnostic." *International Journal of Pharmaceutics* 564 (2019): 308–317. doi:10.1016/j.ijpharm.2019.04.055.

18. Qin, Junjie, Bohua Dong, Rongjie Gao, Ge Su, Jiwei Han, Xue Li, Wei Liu, Wei Wang, and Lixin Cao. "Water-soluble silica-coated ZnS: Mn nanoparticles as fluorescent sensors for the detection of ultratrace copper (II) ions in seawater." *Analytical Methods* 9, no. 2 (2017): 322–328. doi:10.1039/C6AY02486F.

19. Bogireddy, Naveen Kumar Reddy, Victor Barba, and Vivechana Agarwal. "Nitrogen-doped graphene oxide dots-based "Turn-OFF" H_2O_2, Au (III), and "Turn-OFF–ON" Hg (II) sensors as logic gates and molecular keypad locks." *ACS Omega* 4, no. 6 (2019): 10702–10713. doi:10.1021/acsomega.9b00858.

20. Cheng, Chaoge, Malcolm Xing, and Qilin Wu. "A universal facile synthesis of nitrogen and sulfur co-doped carbon dots from cellulose-based biowaste for fluorescent detection of Fe^{3+} ions and intracellular bioimaging." *Materials Science and Engineering: C* 99 (2019): 611–619. doi:10.1016/j.msec.2019.02.003.

21. Xu, Quan, Miaoran Zhang, Yao Liu, Wei Cai, Wenjing Yang, Ziying He, Xiuli Sun, Yan Luo, and Fang Liu. "Synthesis of multi-functional green fluorescence carbon dots and their applications as a fluorescent probe for Hg^{2+} detection and zebrafish imaging." *New Journal of Chemistry* 42, no. 12 (2018): 10400–10405. doi:10.1039/C8NJ01639A.

22. Li, Rong Sheng, Jia Hui Liu, Tong Yang, Peng Fei Gao, Jian Wang, Hui Liu, Shu Jun Zhen, Yuan Fang Li, and Cheng Zhi Huang. "Carbon quantum dots–europium (III) energy transfer architecture embedded in electrospun nanofibrous membranes for fingerprint security and document counterspy." *Analytical Chemistry* 91, no. 17 (2019): 11185–11191. doi:10.1021/acs.analchem.9b01936.

23. Guo, Yongming, Yuzhi Chen, Fengpu Cao, Lijuan Wang, Zhuo Wang, and Yumin Leng. "Hydrothermal synthesis of nitrogen and boron doped carbon quantum dots with yellow-green emission for sensing Cr (VI), anti-counterfeiting and cell imaging." *RSC Advances* 7, no. 76 (2017): 48386–48393. doi:10.1039/C7RA09785A.

24. Fan, Chonghui, Kelong Ao, Pengfei Lv, Jiancheng Dong, Di Wang, Yibing Cai, Qufu Wei, and Yang Xu. "Fluorescent nitrogen-doped carbon dots via single-step synthesis applied as fluorescent probe for the detection of Fe^{3+} ions and anti-counterfeiting inks." *Nano* 13, no. 08 (2018): 1850097. doi:10.1142/S1793292018500972.

25. Liu, Xiangjun, Nan Zhang, Tao Bing, and Dihua Shangguan. "Carbon dots based dual-emission silica nanoparticles as a ratiometric nanosensor for Cu^{2+}." *Analytical Chemistry* 86, no. 5 (2014): 2289–2296. doi:10.1021/ac404236y.

26. Zhang, Li, Zhao-Wu Wang, Sai-Jin Xiao, Dong Peng, Jia-Qing Chen, Ru-Ping Liang, Jun Jiang, and Jian-Ding Qiu. "Fluorescent molybdenum oxide quantum dots and Hg^{2+} synergistically accelerate cobalt porphyrin formation: A new strategy for trace Hg^{2+} analysis." *ACS Applied Nano Materials* 1, no. 4 (2018): 1484–1491. doi:10.1021/acsanm.7b00351.

27. Peng, Dong, Li Zhang, Ru-Ping Liang, and Jian-Ding Qiu. "Rapid detection of mercury ions based on nitrogen-doped graphene quantum dots accelerating formation of manganese porphyrin." *ACS Sensors* 3, no. 5 (2018): 1040–1047. doi:10.1021/acssensors.8b00203.

28. Patir, Khemnath, and Sonit Kumar Gogoi. "Facile synthesis of photoluminescent graphitic carbon nitride quantum dots for Hg^{2+} detection and room temperature phosphorescence." *ACS Sustainable Chemistry & Engineering* 6, no. 2 (2018): 1732–1743. doi:10.1021/acssuschemeng.7b03008.

29. Gogoi, Neelam, Mayuri Barooah, Gitanjali Majumdar, and Devasish Chowdhury. "Carbon dots rooted agarose hydrogel hybrid platform for optical detection and separation of heavy metal ions." *ACS Applied Materials & Interfaces* 7, no. 5 (2015): 3058–3067. doi:10.1021/am506558d.

30. Lu, Shuaimin, Di Wu, Guoliang Li, Zhengxian Lv, Zilin Chen, Lu Chen, Guang Chen, Lian Xia, Jinmao You, and Yongning Wu. "Carbon dots-based ratiometric nanosensor for highly sensitive and selective detection of mercury (II) ions and glutathione." *RSC Advances* 6, no. 105 (2016): 103169–103177. doi:10.1039/C6RA21309J.

31. Kuo, Shih-Yu, Hsiang-Hau Li, Pei-Jing Wu, Chuan-Pin Chen, Ya-Chi Huang, and Yang-Hsiang Chan. "Dual colorimetric and fluorescent sensor based on semiconducting polymer dots for ratiometric detection of lead ions in living cells." *Analytical Chemistry* 87, no. 9 (2015): 4765–4771. doi:10.1021/ac504845t.

32. Gore, Anil H., Dattatray B. Gunjal, Mangesh R. Kokate, Vasanthakumaran Sudarsan, Prashant V. Anbhule, Shivajirao R. Patil, and Govind B. Kolekar. "Highly selective and sensitive recognition of cobalt (II) ions directly in aqueous solution using carboxyl-functionalized CdS quantum dots as a naked eye colorimetric probe: applications to environmental analysis." *ACS Applied Materials & Interfaces* 4, no. 10 (2012): 5217–5226. doi:10.1021/am301136q.

33. Zhao, Ting, Xiandeng Hou, Ya-Ni Xie, Lan Wu, and Peng Wu. "Phosphorescent sensing of Cr^{3+} with protein-functionalized Mn-doped ZnS quantum dots." *Analyst* 138, no. 21 (2013): 6589–6594. doi:10.1039/C3AN01213A.

34. Jaiswal, Amit, Siddhartha Sankar Ghsoh, and Arun Chattopadhyay. "Quantum dot impregnated-chitosan film for heavy metal ion sensing and removal." *Langmuir* 28, no. 44 (2012): 15687–15696. doi:10.1021/la3027573.

35. Fu, Zheng, and Fengling Cui. "Thiosemicarbazide chemical functionalized carbon dots as a fluorescent nanosensor for sensing Cu^{2+} and intracellular imaging." *RSC Advances* 6, no. 68 (2016): 63681–63688. doi:10.1039/C6RA10168B.

36. Ke, Jun, Xinyong Li, Qidong Zhao, Yang Hou, and Junhong Chen. "Ultrasensitive quantum dot fluorescence quenching assay for selective detection of mercury ions in drinking water." *Scientific Reports* 4, no. 1 (2014): 1–6. doi:10.1038/srep05624.

37. Xiao, Meng, Zhonggang Liu, Ningxia Xu, Lelun Jiang, Mengsu Yang, and Changqing Yi. "A smartphone-based sensing system for on-site quantitation of multiple heavy metal ions using fluorescent carbon nanodots-based microarrays." *ACS Sensors* 5, no. 3 (2020): 870–878. doi:10.1021/acssensors.0c00219.

38. Ibrahim, Izwaharyanie, Hong Ngee Lim, Osama K. Abou-Zied, Nay Ming Huang, Pedro Estrela, and Alagarsamy Pandikumar. "Cadmium sulfide nanoparticles decorated with au quantum dots as ultrasensitive photoelectrochemical sensor for selective detection of copper (II) ions." *The Journal of Physical Chemistry C* 120, no. 39 (2016): 22202–22214. doi:10.1021/acs.jpcc.6b06929.

39. Delices, Annette, Davina Moodelly, Charlotte Hurot, Yanxia Hou, Wai Li Ling, Christine Saint-Pierre, Didier Gasparutto, Gilles Nogues, Peter Reiss, and Kuntheak Kheng. "Aqueous synthesis of DNA-functionalized near-infrared $AgInS_2$/ZnS core/shell quantum dots." *ACS Applied Materials & Interfaces* 12, no. 39 (2020): 44026–44038. doi:10.1021/acsami.0c11337.

40. Yuan, Chao, Bianhua Liu, Fei Liu, Ming-Yong Han, and Zhongping Zhang. "Fluorescence "turn on" detection of mercuric ion based on bis (dithiocarbamato) copper (II) complex functionalized carbon nanodots." *Analytical Chemistry* 86, no. 2 (2014): 1123–1130. doi:10.1021/ac402894z.

41. Huang, Dawei, Chenggang Niu, Min Ruan, Xiaoyu Wang, Guangming Zeng, and Canhui Deng. "Highly sensitive strategy for Hg^{2+} detection in environmental water samples using long lifetime fluorescence quantum dots and gold nanoparticles." *Environmental Science & Technology* 47, no. 9 (2013): 4392–4398. doi:10.1021/es302967n.

42. Boobalan, T., M. Sethupathi, N. Sengottuvelan, Ponnuchamy Kumar, P. Balaji, Balázs Gulyás, Parasuraman Padmanabhan, Subramanian Tamil Selvan, and A. Arun. "Mushroom-derived carbon dots for toxic metal ion detection and as antibacterial and anticancer agents." *ACS Applied Nano Materials* 3, no. 6 (2020): 5910–5919. doi:10.1021/acsanm.0c01058.

43. Ren, Qiaoli, Lu Ga, and Jun Ai. "Rapid synthesis of highly fluorescent nitrogen-doped graphene quantum dots for effective detection of ferric ions and as fluorescent ink." *ACS Omega* 4, no. 14 (2019): 15842–15848. doi:10.1021/acsomega.9b01612.

44. Jin, Li-Hua, and Chang-Soo Han. "Ultrasensitive and selective fluorometric detection of copper ions using thiosulfate-involved quantum dots." *Analytical Chemistry* 86, no. 15 (2014): 7209–7213. doi:10.1021/ac501515f.

45. Jin, Yao, Wei Duan, Fangjie Wo, and Jianmin Wu. "Two-dimensional fluorescent strategy based on porous silicon quantum dots for metal-ion detection and recognition." *ACS Applied Nano Materials* 2, no. 10 (2019): 6110–6115. doi:10.1021/acsanm.9b01647.

46. Dutta Chowdhury, Ankan, and Ruey-an Doong. "Highly sensitive and selective detection of nanomolar ferric ions using dopamine functionalized graphene quantum dots." *ACS Applied Materials & Interfaces* 8, no. 32 (2016): 21002–21010. doi:10.1021/acsami.6b06266.

47. Wang, Shen, Jie Yu, Pingnan Zhao, Siyao Guo, and Song Han. "One-Step synthesis of water-soluble CdS quantum dots for silver-ion detection." *ACS Omega* 6, no. 10 (2021): 7139–7146. doi:10.1021/acsomega.1c00162.

48. Fan, Qin, Jinhua Li, Yuhua Zhu, Zilu Yang, Tao Shen, Yizhong Guo, Lihua Wang, Tao Mei, Jianying Wang, and Xianbao Wang. "Functional carbon quantum dots for highly sensitive graphene transistors for Cu^{2+} ion detection." *ACS Applied Materials & Interfaces* 12, no. 4 (2020): 4797–4803. doi:10.1021/acsami.9b20785.

49. Zhang, Yating, Keke Li, Shaozhao Ren, Yongqiang Dang, Guoyang Liu, Ruizhe Zhang, Kaibo Zhang, Xueying Long, and Kaili Jia. "Coal-derived graphene quantum dots produced by ultrasonic physical tailoring and their capacity for Cu (II) detection." *ACS Sustainable Chemistry & Engineering* 7, no. 11 (2019): 9793–9799. doi:10.1021/acssuschemeng.8b06792.

50. Lou, Ying, Jianying Ji, Aimiao Qin, Lei Liao, Ziyuan Li, Shuoping Chen, Kaiyou Zhang, and Jun Ou. "Cane molasses graphene quantum dots passivated by PEG functionalization for detection of metal ions." *ACS Omega* 5, no. 12 (2020): 6763–6772. doi:10.1021/acsomega.0c00098.

51. Chu, Hongtao, Dong Yao, Jiaqi Chen, Miao Yu, and Liqiang Su. "Detection of Hg^{2+} by a dual-fluorescence ratio probe constructed with rare-earth-element-doped cadmium telluride quantum dots and fluorescent carbon dots." *ACS Omega* (2021). doi:10.1021/acsomega.1c00263.

52. Chan, Yang-Hsiang, Jixin Chen, Qingsheng Liu, Stacey E. Wark, Dong Hee Son, and James D. Batteas. "Ultrasensitive copper (II) detection using plasmon-enhanced and photo-brightened luminescence of CdSe quantum dots." *Analytical Chemistry* 82, no. 9 (2010): 3671–3678. doi:10.1021/ac902985p.

53. Zhang, Diwei, Yan Xu, Quanlin Liu, and Zhiguo Xia. "Encapsulation of $CH_3NH_3PbBr_3$ perovskite quantum dots in MOF-5 microcrystals as a stable platform for temperature and aqueous heavy metal ion detection." *Inorganic Chemistry* 57, no. 8 (2018): 4613–4619. doi:10.1021/acs.inorgchem.8b00355.

54. Ding, Nan, Donglei Zhou, Gencai Pan, Wen Xu, Xu Chen, Dongyu Li, Xiaohui Zhang, Jinyang Zhu, Yanan Ji, and Hongwei Song. "Europium-doped lead-free $Cs_3Bi_2Br_9$ perovskite quantum dots and ultrasensitive Cu^{2+} detection." *ACS Sustainable Chemistry & Engineering* 7, no. 9 (2019): 8397–8404. doi:10.1021/acssuschemeng.9b00038.

55. Chen, Pin-Ru, Kuo-Yang Lai, Chang-Wei Yeh, and Hsueh-Shih Chen. "Aminophosphine-based InP quantum dots for the detection of Zn^{2+} and Cd^{2+} ions in water." *ACS Applied Nano Materials* (2021). doi:10.1021/acsanm.1c00336.

56. Dong, Yongqiang, Ruixue Wang, Geli Li, Congqiang Chen, Yuwu Chi, and Guonan Chen. "Polyamine-functionalized carbon quantum dots as fluorescent probes for selective and sensitive detection of copper ions." *Analytical Chemistry* 84, no. 14 (2012): 6220–6224. doi:10.1021/ac3012126.

57. Qiang, Ruibin, Weiming Sun, Kaiming Hou, Zhangpeng Li, Jinyun Zhang, Yong Ding, Jinqing Wang, and Shengrong Yang. "Electrochemical trimming of graphene oxide affords graphene quantum dots for Fe^{3+} detection." *ACS Applied Nano Materials* (2021). doi:10.1021/acsanm.1c00621.

58. Ruedas-Rama, Maria Jose, and Elizabeth AH Hall. "Azamacrocycle activated quantum dot for zinc ion detection." *Analytical Chemistry* 80, no. 21 (2008): 8260–8268. doi:10.1021/ac801396y.

59. Chu, Hongtao, Dong Yao, Jiaqi Chen, Miao Yu, and Liqiang Su. "Double-emission ratiometric fluorescent sensors composed of rare-earth-doped ZnS quantum dots for Hg^{2+} detection." *ACS Omega* 5, no. 16 (2020): 9558–9565. doi:10.1021/acsomega.0c00861.

9 Nanoporous Membrane-Based Plasmonic Nanosensors

9.1 INTRODUCTION

Agricultural planting, industrial production, disease diagnosis, and environmental detection are extensively utilized in numerous human life and social development areas as sensors [1]. Sensors may be categorized into three types consistent with the variations of ecological issues: chemical sensors toward heavy metal ions, metal cations, gas/vapor, pH, and lethal biological composites; physical sensors toward humidity, strain/pressure, and temperature; and biosensors toward pathogens and biomolecules [2–5]. Presently, the traditional instrument analysis devices still show some disadvantages, such as professional operators, complicated operation steps, and expensive instruments, though they benefit from professionalism and sensitivity. The progress of suitable, green, multifunctional, rapid, exceedingly selective, and sensitive sensors is urgently required. The membrane materials have stimulated an extensive concentration of researchers, owing to their many advantages of flexible, real-time, and portable detection. So far, membrane materials are primarily organic substrates and inorganic matrices, for example, ceramics and metal oxides [6–8]. In commercial applications, biological substances show the benefits of large scale, easy processing, and low cost.

In contrast, inorganic mediums have the drawbacks of tough processing, costly, and small-scale commercial usage. In this regard, the cellulose-based membranes own several benefits, which are widely utilized for the detection fields as a new green and organic substrate of outstanding compatibility, biodegradability, and ecofriendly [2]. Various methods such as solution regeneration, vacuum filtration, phase inversion, solution casting, electrospinning, and other preparation methods have been utilized to develop cellulose-based membranes for sensing purposes [9–11]. The cellulose membranes are prepared using cellulose nanofibers, cellulose acetate, triacetyl cellulose, ethyl-cellulose, sodium carboxymethyl cellulose, cellulose nanocrystals, and nitrocellulose the raw materials [12–21]. Through chemical-physical approaches, metal nanowires, graphene-based membranes, quantum dots, carbon dots, fluorophores, CNT, antigen-antibodies, and organic dyes into cellulose membranes are excellent detection performances be attained [22–30]. To the extent that for sensing applications, the cellulose-based membranes show wide-ranging applications. Here, we comprise the current advances in the sensing field of pollutants using cellulose-based membranes and other

DOI: 10.1201/9781003128281-9

changed nanoporous membranes as biosensors, chemical sensors, and physical sensors. The development prospects and challenges of these functionalized membrane-based sensors are briefly discussed.

9.2 APPLICATIONS OF NANOPOROUS MEMBRANE FOR TOXIC METAL ION SENSING

For heavy metal ions sensing applications, the poly(5-sulfo-1-aminoanthraquinone) NPs have been fabricated through the oxidative chemical polymerization of 5-sulfo-1-aminoanthraquinone. Several polymerization parameters were steadily considered predicting advanced structure, multifunctionalities, and synthetic yield of the target NPs, for example, acid species, oxidant species, oxidant/monomer ratio, acid concentration, polymerization time, and temperature. Various analytical instruments have been utilized for the determination of size distribution, molecular structure, properties, and morphology of the NPs, for example, fluorescence spectroscopies, UV–vis, and IR, and for element analyses, simultaneous thermo-gravimetric analysis (TGA), laser particle analyzer, MALDI-MS, AFM, XRD, and FESEM. The fabrication of the NPs with large π-conjugation, a clean surface, intrinsic semiconductivity, narrow size distribution, inherent self-stability, and blue fluorescence is required. The aqueous $HClO_4$ and K_2CrO_4 oxidant are an optimum mixture without any external stabilizer attributed to macromolecular chains of many negatively charged sulfonic groups. Significantly, the NPs show an excellent removal percentage of respective mercury and lead ions with an exclusive synergic combination of –Nd, –NH, –SO_3H, and –NH_2, functional groups with $115.15\,m^{-2}g^{-1}$ specific area. These NPs showed the RF of 99.8% and 99.6%, respectively, for the mercury and Pb^{2+} ions at initial concentrations of $200\,mg\,L^{-1}$. Together with removing these metal ions, the PSA solution can be utilized as an innovative fluorescent chemosensor. The developed chemosensor showed ultrahigh sensitivity and high selectivity toward Pb^{2+} ions owing to its robust anti-interference in mixed metal ion solution and very greater LOD at $1.0 \times 10^{-10}\,M$ [31].

Similarly, the crosslinked chitosan thin film was used as an optical sensor for HMs detection based on the SPR approach. For the fabrication of the sensor, thru a homogeneous reaction of intermediate chitosan with glutaraldehyde as a crosslinking agent, the crosslinked chitosan solution was made in an aqueous acetic acid solution, which was further deposited through spin coating on a gold film. Before and after interaction with a diverse concentration of Pb^{2+} ion, the optical possessions of chitosan thin film (crosslinked) had been attained thru fitting using the SPR method in a range of 0.5–100 ppm. On increasing the concentration of Pb^{2+} ions in the NP solution, the resonance angle was directly proportional to the Pb^{2+} ion concentration and shifted toward a lower value. To confirm the chemical interactions, intricate among the thin film and Pb^{2+} ion and bonding of the thin film and chemical elements before adsorption had been employed using XPS [32].

In another study, for the detection of zinc (Zn^{2+}) ion, novel active nanolayers have been synthesized using the SPR system. Using innovative active nanolayers, that is, chitosan and chitosan–tetrabutyl thiuram disulfide (chitosan–TBTDS), the gold surface was modified, which was further used for the SPR system. On the gold surface, the spin coating method has been used to synthesize chitosan active layers and chitosan–TBTDS. Aqueous Zn^{2+} ions were detected with improved sensitivity through the TBTDS system by monitoring the SPR signal. The resonance angle shift is directly proportional to the Zn^{2+} ion concentration in aqueous media for both active nanolayers. For chitosan–TBTDS active nanolayer, the higher resonance angle shift was attained owing to specific binding of Zn^{2+} ions with TBTDS. The lowest LOD of chitosan–TBTDS active nanolayer toward Zn^{2+} ions found up to 0.1 mg L^{-1} confirms the improved sensitivity of the detection probe [33].

A detection system has been fabricated by varying pendant ionic side chains, mutual poly(p-pheynylene ethynylene) (PPE) backbone, and four anionic conjugated polyelectrolytes (CPEs). In response to the addition of diverse metals, the key factors are the repeat unit pattern, composition of the ionic side chain, and conjugation length, affecting the fluorescence patterns of CPE polymers. In this study, according to the Environmental Protection Agency (EPA), the presence of the eight metal ions considered as water pollutants, for example, Hg^{2+}, Pb^{2+}, Cr^{3+}, Fe^{3+}, Mn^{2+}, Cu^{2+}, Co^{2+}, and Ni^{2+}. The developed four-PPE sensor array was transformed into canonical scores and analyzed and recorded for each metal ion using linear discrimination analysis (LDA). This method allowed apparent differences among metals using both three-dimensional and 2D graphs and response patterns of fluorescence intensity. Especially in distinct aqueous solutions at 100 nM, the array can willingly distinguish among eight toxic metal ions. The developed four-PPE sensors also offer a practical application in blind samples of aqueous Hg^{2+} and Pb^{2+} ions within a specific concentration range [34].

For the detection and fast collection of food pollutants, combining solid-phase extraction (SPE) through SERS, a disordered silver nanowires membrane was utilized. A solvothermal polyol process has been used to prepare silver nanowires colloid solution followed by the fabrication of membrane. Filter membrane acquired silver nanowires to focus liquid-phase analytes in less than 10s in 5 mL. For the study of food safety toxins, the membrane comprised the applications of SERS and SPE methods. The usage of the extraction membrane-active toward the SERS technique shortened the time of analysis through the elimination of the elution procedure. Under continuous laser irradiation, it has been demonstrated that the synthesized membrane had high temporal stability and good uniformity. Based on a flow-through method, quantitative and qualitative detection of melamine and phorate has been done. In the 2.5–100 mg mL^{-1} of concentration range for melamine and 2.5–10 mg mL^{-1} for phorate, the characteristic SERS intensity showed an excellent linear relationship plotted against concentrations of melamine and phorate [35].

For the fast removal and simultaneous detection with multiple signal amplification of Hg^{2+} ions, the fabrication of multifunction system leftovers is a prime challenge nowadays. Therefore, for fast removal and dual signal amplification detection of Hg^{2+} ions, the filter membrane of AuNPs was utilized for the first time as adsorbents and chemosensors. The synthesis of gold amalgam has been done by using colorimetric and fluorescent sensing techniques. By changing the concentration of Hg^{2+} ions deposited on AuNPs of 13 nm, the orange fluorescence and red color of RhB progressively altered to the colorless solution when the gold amalgam catalyzed the reduction of RhB. The LOD of the colorimetric assay and fluorescence assay was found to be 2.54 and 1.16 nM for Hg^{2+}, correspondingly. Fascinatingly, the color and fluorescence of RhB can be improved when the above colorless reaction solution was kept in the air for about 2 h. A recyclable paper-based sensor to recognize Hg^{2+} ions in aqueous samples has been developed to apply the above optical phenomenon by incorporating the RhB dye and Au NPs on filter paper. In yellow river water and tap water, the AuNPs incorporated filter membrane showed fast removal of Hg^{2+} ions, having 99% RF [36].

Similarly, in another study, for detecting Pb^{2+} ions, the chitosan graphene oxide (CS-GO) SPR sensor incorporated with a multi-metallic layer of Au-Ag-Au nanostructure has been employed SPR method. The sensor competence in terms of repeatability, S/N ratio, and sensitivity is analyzed by measurements of SPR. Various analytical tools, such as XPS, FESEM, AFM, XRD, and Raman spectroscopy, have been utilized to characterize the nanostructure layers. The developed nanostructure has augmented the SPR angle shift up to 3.5° because of the improved evanescent field at the sensing layer-analyte interface. The prodigious repeatability of the sensor is also the advantage of the stable multi-metallic nanostructure. Toward the Pb^{2+} ion solution, the trimetallic CS-GO SPR sensor offers a virtuous response in its 5 ppm concentration, which is further confirmed thru 0.92 of the SNR value for Pb^{2+} ion solution of 5 ppm concentration. The trimetallic CS-GO SPR sensor showed reliability and robust capability to detect Pb^{2+} ions consistent with the standardized Pb^{2+} safety level for wastewater [37].

For the detection Pb^{2+} ions in the water sample, poly(m-phenylenediamine-co-aniline-2-sulfonic acid) (poly(mPD-co-ASA)) specific receptors have been developed. Exceedingly sensitive optochemical sensor based on LSPR, which was further functionalized with self-assembly gold nanoislands (SAMAuNIs) copolymer NPs. The surface density of the copolymer, size of copolymer NPs, and ASA:mPD monomers mole ratio are the three optimized aspects of the copolymer receptor. It is demonstrated that the copolymer solution diluted 200 times pays to the utmost operative protocol for functionalization, and the ASA:mPD copolymer of 5:95 mole ratio displays the best Pb^{2+}- sensing performance with a size less than 100 nm. The ensuing poly(mPD-co-ASA)-functionalized LSPR sensor in the 0.011–5,000 ppb linear dynamic range showed the LOD up to 0.011 ppb toward sensing Pb^{2+} ions in a water sample. Furthermore, in other common anions and metallic cations, the sensing system displays robust selectivity toward Pb^{2+} ions. For rapid drinking water inspection using simple LSPR instrumentation, the anticipated AuNIs functionalized functional copolymer is found to offer outstanding Pb^{2+}-sensing capability [38].

Similarly, owing to its potential advantages in numerous fields, a 2D graphene-like material, that is, molybdenum disulfide (MoS_2), is prodigious fascinating consideration. Therefore, a self-assembled MoS_2@TMPyP nanocomposite fabricated comprising 5,10,15,20-tetrakis(1-methyl-4- pyridinio)porphyrintetra(p-toluene-sulfonate) (TMPyP) and MoS_2 nanosheets. This developed nanocomposite was used as a detection probe for selective, rapid, low-cost detection of Cd^{2+} ions. In observed UV–vis absorption spectra, a new Soret band appeared at 442 nm because of the coordination of Cd^{2+} ions with TMPyP of the MoS_2@TMPyP. The coordination rates are 200 times faster in the presence of MoS_2 among TMPyP, and Cd^{2+} ions contrast to nonattendance of MoS_2, which is significantly enhanced with the help of MoS_2 from 72 h to 20 min. The developed probe showed the lowest LOD at 7.2×10^{-8} mol L^{-1} toward the Cd^{2+} ions. The binding actions among the MoS_2 nanosheets and cationic TMPyP were verified thru various control experiments and molecular dynamics simulation. For driving TMPyP, the electrostatic interaction was the key force after an enhanced complexation of TMPyP and Cd^{2+} rousing around the MoS_2 surface. For eliminating Cd^{2+} in water, the developed nanocomposite of MoS_2@TMPyP could be utilized. For high concentrations of Cd^{2+} ions, the RF can reach 91% of the MoS_2@TMPyP nanocomposite. The developed method offers an innovative perception for eliminating Cd^{2+} metal ions and their detection in water [39].

In another study, to detect heavy metals selectively, ion-imprinted polymers (IIPs) have fascinated vast consideration. For the imprinted cavities selectivity improvement, 8-hydroxyquinoline-grafted gelatin was introduced as a biomolecular monomer with diverse functional groups. A 3D IIPs was developed with rapid kinetics, the imprinted factor of 2.9 with strong selectivity, high adsorption capacity up to 235.7 mg g^{-1}. The development of the hydrogel film using a simple crosslinking process followed the swelling/folding was demonstrated based on its film-forming and swelling properties. Diverse morphologies of IIPs can be fabricated based on the diverse swelling vessels to fulfill the necessities of practical application. Therefore, to determine Pb^{2+} ions in surface water and drinking water, the spectrophotometric coupled IIPs extraction was useful, which showed 0.2 ng mL^{-1} of the LOD [40].

Inherently poor reactivity and low conductivity limit various semiconductors for electrochemical detection. Therefore, to make them unstable, expensive, and complex carbon- and metal-based sensors, some changes in the properties of semiconductors are essential. In this, the useful concept for semiconductors is surface-electronic-state-modulation. With no modifications, the pure semiconductors are allowed to be directly accessible for the electrochemical sensing of heavy metals with ultra-sensitivity. The developed TiO_2 nanosheet of defective single-crystalline (001) showed the LOD up to 0.017 μM with 270.83 μA $μM^{-1}cm^{-2}$ of high sensitivity showed outstanding electrochemical capacity toward Hg^{2+} ions. The consumption limit is 0.03 μM, according to WHO. The surface electronic-state modulation process extends the electrochemical detection of pure semiconductors to detect the heavy metal ions up to the atomic level [41].

Similarly, in flexible electronic devices, graphene-coated plastic substrates are regularly used, for example, polyethylene terephthalate (PET). Therefore, to detect heavy metals selectively, a novel advantage of the nanoporous PET membrane is based on the graphene-coated materials used in an ion exchange manner.

Swift heavy ions irradiation is utilized to perforate PET substrate and graphene. After chemical etching, to support graphene nanopores, conical holes in the PET can generate graphene nanopores using the carboxyl groups. A nanoporous graphene monolayer membrane is fruitfully fabricated with a PET substrate to examine its ionic selective separation. In the driving solution, the ions permeation ratio sturdily relies on the H^+ concentration and temperature. The permeation ratio of ions can be increased thru an electric field through nanoporous graphene. It hinders the separation of ions selectively. At the density functional theory (DFT) level, the graphene nanopore structure is resolved with -COOH functional groups. The results showed that the asymmetric structure could be attributed to the ion exchange of the nanopore with carboxyl groups among protons and heavy metals. These outcomes would be helpful in designing the membrane separation materials made from graphene with effective offline and online bulk separation [42].

In the practical application and basic research of nanofluidic devices, nanochannels play an important role in the transport of ionic species. Therefore, on the surface of porous anodized aluminum oxide (PAA) membrane, a visualized CdSe@ZIF-8/PAA nanochannel membrane was fabricated using precursor solvents, that is, zinc nitrate, 2-methylimidazole, and CdSe QDs. Here, in situ growth has synthesized zeolite imidazole skeleton (ZIF-8) and CdSe quantum dots (CdSe QDs). The ZIF-8 is a metal-organic framework used as a microporous material that possesses a strong metal adsorption capacity. Furthermore, CdSe quantum dots partake in fluorescent possessions. Thru the interaction among Se and S atoms and Cu^{2+} ions, the quenching of fluorescence intensity is responsible for detecting the Cu^{2+} ions on the nanochannel membrane. The 5 V of direct potential was used to attain Cu^{2+} enhancement at the nanochannel interface resulted in the change in fluorescence. The CdSe@ZIF-8/PAA nanochannel membrane showed Cu^{2+} ions at 0.01 pM to 1 µM of concentration range. The developed probe showed the lowest LOD at 4 fM using the improvement of nanochannel [43]. The schematic representation of the sensing mechanism has been shown in Figure 9.1.

Similarly, for significant water-related ecological advantages, current developments have assisted in the fabrication of MoS_2 nanofilms. Though, for heavy metal detection, a MoS_2 nanofilm-coated sensor has been utilized in aqueous samples. A vertically aligned novel 2D MoS_2 (edge exposed) nanofilm for in situ Pb^{2+} ion detection has been utilized. At -0.45 V vs Ag/AgCl, the fabricated sensor showed an outstanding sensing application in a linear range of 0–20 ppb toward Pb^{2+} ions using square wave anodic stripping voltammetry (SWASV). The synthesized sensing probe showed 0.3 ppb of LOD in a tap water sample. In this, 2.6 times higher sensitivity has been established by the vertically aligned 2D MoS_2 layers compared to horizontally aligned 2D MoS_2 (basal plane exposed).

Contrary to the 0.36 and 0.07 eV basal plane, the MoS_2 side edge shows much higher adsorption energy (4.11 eV) of Pb^{2+} ions using DFT calculations. The center for a bandgap of vertical MoS_2 reduced Pb^{2+} ions and was more advanced than the reduction potential level of the $Pb^{2+} \rightarrow Pb$. Conclusively, for detecting Pb^{2+} in a real drinking water sample, the newly developed 2D MoS_2 sensor aligned vertically showed outstanding capability with good reliability [44].

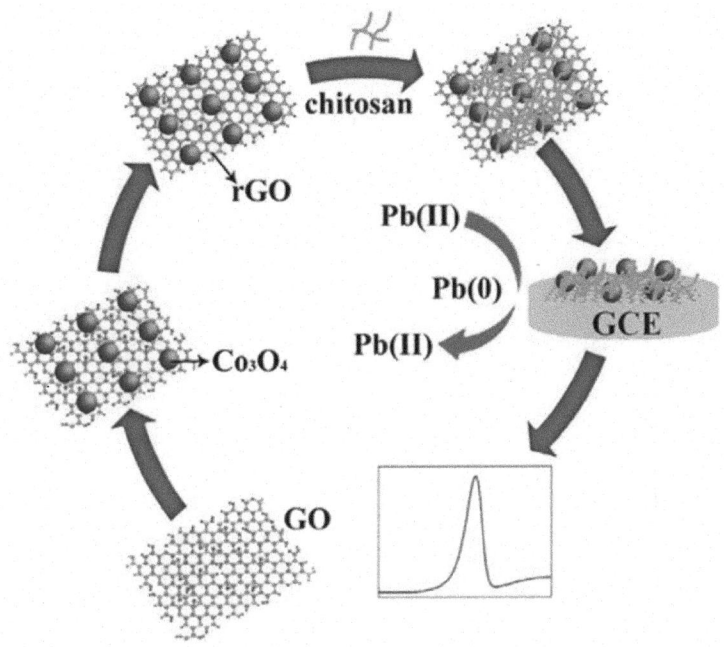

FIGURE 9.1 Diagrammatic representation of nanoporous membrane for the detection of Pb²⁺ metal ion. (Reprinted with permission from Ref. [28]. Copyright 2021 American Chemical Society.)

Because of the environmental pollution because of toxic metals, detection of Hg^{2+} ions is a beneficial interest and fundamental problem at ppb concentration levels. Toward the detection of Hg^{2+} ions, widespread research has been focused mainly on this area. A worldwide detection probe comprising all three oxidation states of Hg^{2+}, that is, 0, +1, and +2, is highly toxic, whereas confirming linearity of response, selectivity, high sensitivity, and enabling in situ deployment would be a significant addition to ex-situ. The thin film of poly(vinyl alcohol) embedded silver nanoparticle (Ag-PVA) was a selective, efficient and fast sensor toward Hg, Hg_2^{2+}, and Hg^{2+} in water, which showed the lowest detection limit at one ppb. The 10 ppb to 1 ppm is the linear concentration range for sensor response. The sensing method was highly selective, which is the exclusive property of the thin film-based sensor, due to the reduction in the SPR absorption upon interaction with the blue shift with Hg^{2+} ions. Most known sensors work only in situ, whereas the thin film sensor could be utilized ex-situ. Investigation of the thin film offers a comprehensive understanding of the sensing systems using spectroscopy and microscopy through the sensing systems [45].

In addition, nonconjugated block copolymers-based light-emitting material has been developed comprising polystyrene sulfonate (PSS) chains. The PSS chains incarceration was found to be a competent technique within nanosized domains to achieve improved fluorescence signals. A high fluorescence QY was attained through altering the types of self-assembled morphologies of PSS-containing block copolymers with a maximum value of 37% in contrast to the lacking organization of PSS homopolymer, which was merely 5% of fluorescence QY attained. Thru rational molecular design, the emission wavelength of the fluorescence process can be tuned. Substantial self-quenching actions were not noted in this material, such as free-standing membranes, thin films, and solutions. Remarkably, the block copolymer electrolytes with light-emitting self-assembled property showed a rapid response time of ≤1 min, inconsequential interfering of mixed metals, and a LOD of parts per billion levels with increased sensitivity for Cu^{2+} metal ions [46].

In a similar study, the polymeric sensor of oligonucleotide-derived thermal-responsive (TBC-P1) nature has been used for Hg^{2+} ion detection. The developed TBC-P1 sensor showed cost-effective and green detection of Hg^{2+} ions via tuning temperature for easy sensor separation and alterable binding among Hg^{2+} ions and oligonucleotides. Using turn-off fluorescence emission, the TBC-P1 sensor showed 0.65 nM of sensing limit with rapid and specific detection properties in water samples toward Hg^{2+} ions. Forming a hydrophobic core, the oligonucleotide-containing thermoresponsive PNIPAM block showed 0.17 nM of LOD, and through warming to yield spherical micelles, it amplified fluorescence signals. Treating the Hg^{2+}-trapped micelles with cysteine (Cys) corresponds to the competition-induced release of this combined sensor TBC-P1 and Hg^{2+} ions, and then recycling polymeric and thermal precipitation. This thermal-responsive polymeric sensor derived from oligonucleotide achieves the goal of improving detection sensitivity and reducing the cost of sensors, which will offer a wide field for recycling sensors [47].

9.3 CONCLUSION AND FUTURE PERSPECTIVES

In this chapter, in the fields of metal cations, non-metal ions, toxic organic compounds, pathogens, and other sensing applications have been comprised of the development of cellulose membranes and their modern advances. The functionalized cellulose membranes can show outstanding sensing performances and specify that cellulose membranes have broad application prospects and great momentums in detecting heavy metal ions. For example, metal NPs, metal nanowires, carbon dots, reduced graphene oxide/graphene oxide, CNT, quantum dots, antigen-antibodies, fluorophores, and organic dyes incorporated thru chemical and physical bonding in cellulose membranes. However, there are still some difficulties arising here:

- In severe surroundings, for example, high temperature, strong acid, strong alkali, a strong solvent, increased ionic strength, increased pressure, and the cellulose membranes enduring steadiness still need to be measured.
- In the sensing fields, intrinsic methodological bottlenecks of sensing sensitivity, selectivity, reversibility, and response time have still occurred.

- The detection response mechanisms based on the theoretic plan, the connection among the properties and structures of cellulose membranes, must be discovered because these are still indistinct.
- For attaining multifunctional concurrent detection of heavy metals, cellulose membrane-based sensors had infrequently been described.
- Though several sensing membranes based on cellulose have been fabricated, several sensing membranes can be commercialized out of them.

Furthermore, future applied applications will cause superior development of recyclable, low-cost, biocompatible, portable, integrated, intelligent, and miniaturized properties. More new applications should also be explored of cellulose sensing membranes in interdisciplinary fields and other fields. To face these challenges of cellulose membrane-based sensors, with continuous efforts, we believe that the road ahead is tortuous, although the future is bright.

REFERENCES

1. Ghoneim, M. T., A. Nguyen, N. Dereje, J. Huang, G. C. Moore, P. J. Murzynowski, and C. Dagdeviren. "Recent progress in electrochemical pH-sensing materials and configurations for biomedical applications." *Chemical Reviews* 119, no. 8 (2019): 5248–5297. doi:10.1021/acs.chemrev.8b00655.
2. Dai, Lei, Yan Wang, Xuejun Zou, Zhirong Chen, Hong Liu, and Yonghao Ni. "Ultrasensitive physical, bio, and chemical sensors derived from 1-, 2-, and 3-D nanocellulosic materials." *Small* 16, no. 13 (2020): 1906567. doi:10.1002/smll.201906567.
3. Zhang, Zhao, Gang Liu, Xinping Li, Sufeng Zhang, Xingqiang Lü, and Yaoyu Wang. "Design and synthesis of fluorescent nanocelluloses for sensing and bioimaging applications." *ChemPlusChem.* (2020). doi:10.1002/cplu.201900746.
4. Fan, Jiang, Sufeng Zhang, Fei Li, and Junwei Shi. "Cellulose-based sensors for metal ions detection." *Cellulose* 27, no. 10 (2020): 5477–5507. doi:10.1007/s10570-020-03158-x.
5. Fan, Jiang, Sufeng Zhang, Yongshe Xu, Ning Wei, Ben Wan, Liwei Qian, and Ye Liu. "A polyethylenimine/salicylaldehyde modified cellulose Schiff base for selective and sensitive Fe^{3+} detection." *Carbohydrate Polymers* 228 (2020): 115379. doi:10.1016/j.carbpol.2019.115379.
6. Lee, Wen Jie, Yueping Bao, Xiao Hu, and Teik-Thye Lim. "Hybrid catalytic ozonation-membrane filtration process with CeO_x and MnO_x impregnated catalytic ceramic membranes for micropollutants degradation." *Chemical Engineering Journal* 378 (2019): 121670. doi:10.1016/j.cej.2019.05.031.
7. Bassyouni, M., M. H. Abdel-Aziz, M. Sh Zoromba, S. M. S. Abdel-Hamid, and Enrico Drioli. "A review of polymeric nanocomposite membranes for water purification." *Journal of Industrial and Engineering Chemistry* 73 (2019): 19–46. doi:10.1016/j.jiec.2019.01.045.
8. Weishaupt, Ramon, Janina N. Zünd, Lukas Heuberger, Flavia Zuber, Greta Faccio, Francesco Robotti, Aldo Ferrari et al. "Antibacterial, cytocompatible, sustainably sourced: Cellulose membranes with bifunctional peptides for advanced wound dressings." *Advanced Healthcare Materials* 9, no. 7 (2020): 1901850. doi:10.1002/adhm.201901850.

9. Mautner, Andreas. "Nanocellulose water treatment membranes and filters: A review." *Polymer International* 69, no. 9 (2020): 741–751. doi:10.1002/pi.5993.

10. Douglass, Eugene F., Huseyin Avci, Ramiz Boy, Orlando J. Rojas, and Richard Kotek. "A review of cellulose and cellulose blends for preparation of bio-derived and conventional membranes, nanostructured thin films, and composites." *Polymer Reviews* 58, no. 1 (2018): 102–163. doi:10.1080/15583724.2016.1269124.

11. Thakur, Vijay Kumar, and Stefan Ioan Voicu. "Recent advances in cellulose and chitosan-based membranes for water purification: A concise review." *Carbohydrate Polymers* 146 (2016): 148–165. doi:10.1016/j.carbpol.2016.03.030.

12. Fu, Wanlin, Yunqian Dai, Xiangyu Meng, Wanlin Xu, Jie Zhou, Zhenguo Liu, Weibing Lu, Shimei Wang, Chaobo Huang, and Yueming Sun. "Electronic textiles based on aligned electrospun belt-like cellulose acetate nanofibers and graphene sheets: Portable, scalable and eco-friendly strain sensor." *Nanotechnology* 30, no. 4 (2018): 045602. doi:10.1088/1361-6528/aaed99.

13. Zhu, Li, Xin Zhou, Yuhang Liu, and Qiang Fu. "Highly sensitive, ultrastretchable strain sensors prepared by pumping hybrid fillers of carbon nanotubes/cellulose nanocrystal into electrospun polyurethane membranes." *ACS Applied Materials & Interfaces* 11, no. 13 (2019): 12968–12977. doi:10.1021/acsami.9b00136.

14. Zhu, Penghui, Yu Liu, Zhiqiang Fang, Yudi Kuang, Yazeng Zhang, Congxing Peng, and Gang Chen. "Flexible and highly sensitive humidity sensor based on cellulose nanofibers and carbon nanotube composite film." *Langmuir* 35, no. 14 (2019): 4834–4842. doi:10.1021/acs.langmuir.8b04259.

15. Li, Jun-jie, Chang-jun Hou, Dan-qun Huo, Cai-hong Shen, Xiao-gang Luo, Huanbao Fa, Mei Yang, and Jun Zhou. "Detection of trace nickel ions with a colorimetric sensor based on indicator displacement mechanism." *Sensors and Actuators B: Chemical* 241 (2017): 1294–1302. doi:10.1016/j.snb.2016.09.191.

16. Li, Jun-jie, Xian-feng Wang, Dan-qun Huo, Chang-jun Hou, Huan-bao Fa, Mei Yang, and Liang Zhang. "Colorimetric measurement of Fe^{3+} using a functional paper-based sensor based on catalytic oxidation of gold nanoparticles." *Sensors and Actuators B: Chemical* 242 (2017): 1265–1271. doi:10.1016/j.snb.2016.09.039.

17. Li, Naixu, Jun Yin, Lingfei Wei, Quanhao Shen, Wei Tian, Jing Li, Yong Chen, Jing Jin, Hongcheng Teng, and Jiancheng Zhou. "Facile synthesis of cellulose acetate ultrafiltration membrane with stimuli-responsiveness to pH and temperature using the additive of F127-b-PDMAEMA." *Chinese Journal of Chemistry* 35, no. 7 (2017): 1109–1116. doi:10.1002/cjoc.201600820.

18. Amin, Alaa S. "Application of a triacetylcellulose membrane with immobilizated of 5-(2′, 4′-dimethylphenylazo)-6-hydroxypyrimidine-2, 4-dione for mercury determination in real samples." *Sensors and Actuators B: Chemical* 221 (2015): 1342–1347. doi:10.1016/j.snb.2015.07.106.

19. Duong, Hong Dinh, Han Lae Kim, and Jong Il Rhee. "Development of colorimetric and ratiometric fluorescence membranes for detection of nitrate in the presence of aluminum-containing compounds." *Sensors* 18, no. 9 (2018): 2883. doi:10.3390/s18092883.

20. Hittini, Waseem, Ayah F. Abu-Hani, N. Reddy, and Saleh T. Mahmoud. "Cellulose-Copper Oxide hybrid nanocomposites membranes for H_2S gas detection at low temperatures." *Scientific Reports* 10, no. 1 (2020): 1–9. doi:10.1038/s41598-020-60069-4.

21. Tang, Rui Hua, Li Na Liu, Su Feng Zhang, Ang Li, and Zedong Li. "Modification of a nitrocellulose membrane with cellulose nanofibers for enhanced sensitivity of lateral flow assays: Application to the determination of *Staphylococcus aureus*." *Microchimica Acta* 186, no. 12 (2019): 1–8. doi:10.1007/s00604-019-3970-z.

22. Lou, Mengna, Ibrahim Abdalla, Miaomiao Zhu, Jianyong Yu, Zhaoling Li, and Bin Ding. "Hierarchically rough structured and self-powered pressure sensor textile for motion sensing and pulse monitoring." *ACS Applied Materials & Interfaces* 12, no. 1 (2019): 1597–1605. doi:10.1021/acsami.9b19238.

23. Zhang, Hao, Xiaohang Sun, Martin A. Hubbe, and Lokendra Pal. "Highly conductive carbon nanotubes and flexible cellulose nanofibers composite membranes with semi-interpenetrating networks structure." *Carbohydrate Polymers* 222 (2019): 115013. doi:10.1016/j.carbpol.2019.115013.

24. Zhang, Hui, Shilong Zhao, Xiuli Wang, Xiaotong Ren, Jiatao Ye, Lihui Huang, and Shiqing Xu. "The enhanced photoluminescence and temperature sensing performance in rare earth doped $SrMoO_4$ phosphors by aliovalent doping: From material design to device applications." *Journal of Materials Chemistry C* 7, no. 47 (2019): 15007–15013. doi:10.1039/c9tc04965g.

25. Zhang, Longfei, Shaoyi Lyu, Qijun Zhang, Yuntao Wu, Chuck Melcher, Stephen C. Chmely, Zhilin Chen, and Siqun Wang. "Dual-emitting film with cellulose nanocrystal-assisted carbon dots grafted $SrAl_2O_4$, Eu^{2+}, Dy^{3+} phosphors for temperature sensing." *Carbohydrate Polymers* 206 (2019): 767–777. doi:10.1016/j.carbpol.2018.11.031.

26. Jiang, Xiangyang, Jian Xia, and Xiaogang Luo. "Simple, rapid, and highly sensitive colorimetric sensor strips from a porous cellulose membrane stained with Victoria blue B for efficient detection of trace Cd (II) in water." *ACS Sustainable Chemistry & Engineering* 8, no. 13 (2020): 5184–5191. doi:10.1021/acssuschemeng.9b07614.

27. Jiang, Yingnan, Xiaojie Zhang, Lizhi Xiao, Ruyue Yan, Jingwei Xin, Chunxia Yin, Yunxiao Jia et al. "Preparation of dual-emission polyurethane/carbon dots thermoresponsive composite films for colorimetric temperature sensing." *Carbon* 163 (2020): 26–33. doi:10.1016/j.carbon.2020.03.013.

28. Tajik, Somayeh, Hadi Beitollahi, Fariba Garkani Nejad, Zahra Dourandish, Mohammad A. Khalilzadeh, Ho Won Jang, Richard A. Venditti, Rajender S. Varma, and Mohammadreza Shokouhimehr. "Recent developments in polymer nanocomposite-based electrochemical sensors for detecting environmental pollutants." *Industrial & Engineering Chemistry Research* 60, no. 3 (2021): 1112–1136. doi:10.1021/acs.iecr.0c04952.

29. Lv, Pengfei, Huimin Zhou, Alfred Mensah, Quan Feng, Keyu Lu, Jieyu Huang, Dawei Li, Yibing Cai, Lucian Lucia, and Qufu Wei. "In situ 3D bacterial cellulose/nitrogen-doped graphene oxide quantum dot-based membrane fluorescent probes for aggregation-induced detection of iron ions." *Cellulose* 26, no. 10 (2019): 6073–6086. doi:10.1007/s10570-019-02476-z.

30. Shankar, Shiv, Amina Baraketi, Sabato D'Auria, Carole Fraschini, Stephane Salmieri, Majid Jamshidian, Marie Christine Etty, and Monique Lacroix. "Development of support based on chitosan and cellulose nanocrystals for the immobilization of anti-Shiga toxin 2B antibody." *Carbohydrate Polymers* 232 (2020): 115785. doi:10.1016/j.carbpol.2019.115785.

31. Huang, Mei-Rong, Shao-Jun Huang, and Xin-Gui Li. "Facile synthesis of polysulfoaminoanthraquinone nanosorbents for rapid removal and ultrasensitive fluorescent detection of heavy metal ions." *The Journal of Physical Chemistry C* 115, no. 13 (2011): 5301–5315. doi:10.1021/jp1099706.

32. Fen, Yap Wing, W. Mahmood Mat Yunus, and Zainal Abidin Talib. "Analysis of Pb^{2+} ion sensing by crosslinked chitosan thin film using surface plasmon resonance spectroscopy." *Optik* 124, no. 2 (2013): 126–133. doi:10.1016/j.ijleo.2011.11.035.

33. Fen, Yap Wing, W. Mahmood Mat Yunus, Zainal Abidin Talib, and Nor Azah Yusof. "Development of surface plasmon resonance sensor for determining zinc ion using novel active nanolayers as probe." *Spectrochimica Acta Part A: Molecular and Biomolecular Spectroscopy* 134 (2015): 48–52. doi:10.1016/j.saa.2014.06.081.

34. Wu, Yi, Ying Tan, Jiatao Wu, Shangying Chen, Yu Zong Chen, Xinwen Zhou, Yuyang Jiang, and Chunyan Tan. "Fluorescence array-based sensing of metal ions using conjugated polyelectrolytes." *ACS Applied Materials & Interfaces* 7, no. 12 (2015): 6882–6888. doi:10.1021/acsami.5b00587.

35. Shi, Yu-E., Limei Li, Min Yang, Xiaohong Jiang, Quanqin Zhao, and Jinhua Zhan. "A disordered silver nanowires membrane for extraction and surface-enhanced Raman spectroscopy detection." *Analyst* 139, no. 10 (2014): 2525–2530. doi:10.1039/C4AN00163J.

36. Chen, Gaosong, Jun Hai, Hao Wang, Weisheng Liu, Fengjuan Chen, and Baodui Wang. "Gold nanoparticles and the corresponding filter membrane as chemosensors and adsorbents for dual signal amplification detection and fast removal of mercury (II)." *Nanoscale* 9, no. 9 (2017): 3315–3321. doi:10.1039/C6NR09638G.

37. Kamaruddin, Nur Hasiba, Ahmad Ashrif A. Bakar, Mohd Hanif Yaacob, Mohd Adzir Mahdi, Mohd Saiful Dzulkefly Zan, and Sahbudin Shaari. "Enhancement of chitosan-graphene oxide SPR sensor with a multi-metallic layers of Au–Ag–Au nanostructure for lead (II) ion detection." *Applied Surface Science* 361 (2016): 177–184. doi:10.1016/j.apsusc.2015.11.099.

38. Qiu, Guangyu, Siu Pang Ng, Xiongyi Liang, Ning Ding, Xiangfeng Chen, and Chi-Man Lawrence Wu. "Label-free LSPR detection of trace lead (II) ions in drinking water by synthetic poly (mPD-co-ASA) nanoparticles on gold nanoislands." *Analytical Chemistry* 89, no. 3 (2017): 1985–1993. doi:10.1021/acs.analchem.6b04536.

39. Yin, Wenyan, Xinghua Dong, Jie Yu, Jun Pan, Zhiyi Yao, Zhanjun Gu, and Yuliang Zhao. "MoS₂-nanosheet-assisted coordination of metal ions with porphyrin for rapid detection and removal of cadmium ions in aqueous media." *ACS Applied Materials & Interfaces* 9, no. 25 (2017): 21362–21370. doi:10.1021/acsami.7b04185.

40. Huang, Kai, Bingbing Li, Feng Zhou, Surong Mei, Yikai Zhou, and Tao Jing. "Selective solid-phase extraction of lead ions in water samples using three-dimensional ion-imprinted polymers." *Analytical Chemistry* 88, no. 13 (2016): 6820–6826. doi:10.1021/acs.analchem.6b01291.

41. Zhou, Wen-Yi, Jin-Yun Liu, Jie-Yao Song, Jin-Jin Li, Jin-Huai Liu, and Xing-Jiu Huang. "Surface-electronic-state-modulated, single-crystalline (001) TiO₂ nanosheets for sensitive electrochemical sensing of heavy-metal ions." *Analytical Chemistry* 89, no. 6 (2017): 3386–3394. doi:10.1021/acs.analchem.6b04023.

42. Li, Zhan, Yanqi Liu, Yang Zhao, Xin Zhang, Lijuan Qian, Longlong Tian, Jing Bai et al. "Selective separation of metal ions via monolayer nanoporous graphene with carboxyl groups." *Analytical Chemistry* 88, no. 20 (2016): 10002–10010. doi:10.1021/acs.analchem.6b02175.

43. Gao, Hongli, Ruikun Sun, Lei He, Zhong-Ji Qian, Chunxia Zhou, Pengzhi Hong, Shengli Sun, Rijian Mo, and Chengyong Li. "In situ growth visualization nano-channel membrane for ultrasensitive copper ion detection under the electric field enrichment." *ACS Applied Materials & Interfaces* 12, no. 4 (2020): 4849–4858. doi:10.1021/acsami.9b21714.

44. Hwang, Jae-Hoon, Md Ashraful Islam, Heechae Choi, Tae-Jun Ko, Kelsey L. Rodriguez, Hee-Suk Chung, Yeonwoong Jung, and Woo Hyoung Lee. "Improving electrochemical Pb²⁺ detection using a vertically aligned 2D MoS₂ nanofilm." *Analytical Chemistry* 91, no. 18 (2019): 11770–11777. doi:10.1021/acs.analchem.9b02382.

45. Ramesh, G. V., and T. P. Radhakrishnan. "A universal sensor for mercury (Hg, HgI, HgII) based on silver nanoparticle-embedded polymer thin film." *ACS Applied Materials & Interfaces* 3, no. 4 (2011): 988–994. doi:10.1021/am200023w.

46. Ahn, Hyungmin, Sung Yeon Kim, Onnuri Kim, Ilyoung Choi, Chang-Hoon Lee, Ji Hoon Shim, and Moon Jeong Park. "Blue-emitting self-assembled polymer electrolytes for fast, sensitive, label-free detection of Cu (II) ions in aqueous media." *Acs Nano* 7, no. 7 (2013): 6162–6169. doi:10.1021/nn402037x.

47. Wang, Lei, Yu Wang, Wenxuan Li, Wenjing Zhi, Yuanyuan Liu, Liang Ni, and Yun Wang. "Recyclable DNA-derived polymeric sensor: Ultrasensitive detection of Hg (II) ions modulated by morphological changes." *ACS Applied Materials & Interfaces* 11, no. 43 (2019): 40575–40584. doi:10.1021/acsami.9b13035.

10 Carbon Nanotubes-Based Plasmonic Nanosensors

10.1 INTRODUCTION

Because of their excellent properties, such as fast electron transfer rate, high surface area, and thermal and outstanding electrical conductivities, CNTs have fascinated ample consideration toward sensing field [1–5]—the bulk properties of CNTs owned by highly ordered CNT arrays. For instance, CNT electrode arrays show a decreased influence from solution resistance and improved mass transport for various applications, which categorize them as outstanding electrochemical transducers [6,7]. In contrast to arbitrarily ordered CNTs as the electrode, the aligned CNT electrode has a high signal-to-noise ratio, reduced influence from the resistance of solution, and upsurged mass sensitivity, corresponding to lower detection limits and much lower background current [8–12]. The edge-plane-like nanotube ends undergo outer-sphere electron transfer, and these are recognized to be accountable for the improved electrochemical response and fast heterogeneous electron transfer rates for redox couples, for example, $Fe(CN)_6^{3-}/Fe(CN)_6^{4-}$ [10,13]. Several CNT electrode structures have been explored and shown to have the lowest detection limits for the heavy metals using stripping voltammetry [14–18]. CNT threads own the applications of CNTs because these are spun from shorter CNTs and avoid the toxicity issues of individual CNTs. Aligning nanotubes in fibers is an effective way to exploit the anisotropic properties of individual CNTs, for example, electrodes for ASV for micro/macro scale usages.

In contrast to CNT arrays and towers, CNT threads have applications of smaller diameter and longer length, ease of handling, good tensile strength, and high electrical conductivity [14]. For the progress of electrochemical sensors, CNT threads also keep the electrocatalytic properties of CNTs and partake a high surface area, making them possibly beneficial. Previously, CNT thread working electrodes attained a LOD up to 1.4 nM for Zn^{2+} heavy metal ions [16]. To comprise the concurrent detection of multiple metals with CNT thread electrodes such as Pb^{2+}, Cu^{2+}, Zn^{2+}, and Cd^{2+}, with limit of detection in the nanomolar range [14]. For applications that require disposable, easily fabricated, and inexpensive electrochemical cells, their easy fabrication with the equipment uses the potentially used CNT thread. Therefore, owing to the above-discussed properties of CNTs, these play a vital role in fluorescence, electrochemical, and colorimetric sensors toward detecting heavy metal ions and other pollutants in biogenic systems and aqueous media.

DOI: 10.1201/9781003128281-10

10.2   DETECTION OF HEAVY METAL IONS USING CARBON NANOTUBES SENSOR

The MWCNTs nanocomposites have been developed to detect Zn^{2+} and Cu^{2+} ions through LBL assembly of titanium carbide ($Ti_3C_2T_x$) changed with gold. To manufacture this novel nanocomposite electrode, it has done the functionalization of the surface of Au electrodes using an LBL drop-coating process with ultrasonication treated $Ti_3C_2T_x$/MWNTs. To further improve detection capacity through an in situ instantaneous target analytes deposition, "green metal" Sb (antimony) was accomplished. For the electrochemical measurement, SWASV has been used for these analytes. Under the optimal experimental conditions, the fabricated sensor showed outstanding detection capacity. The developed probe showed the LOD for Zn^{2+} and Cu^{2+} ions up to 1.5 and 0.1 ppb, correspondingly. In a wide range of concentration, that is, sweat Zn^{2+}: 500–1,500 ppb, and Cu^{2+}: 300–1,500 ppb; urine Zn^{2+}: 200–600 ppb, and Cu^{2+}: 10–500 ppb), the Zn^{2+} and Cu^{2+} ions were fruitfully detected in biofluids, for example, sweat and urine. The developed flexible sensor has additional applications of excellent stability and ultra-repeatability. Therefore, in the future, these applications offer an excellent possibility for the noninvasive smart detection of heavy metals [19].

Similarly, for the detection of Ag^+ ions, through the nitrogen-doped carbon nanotubes (N-CNTs) immobilized with glucose oxidase, an amperometric glucose biosensor was fruitfully synthesized. The amperometric oxidation of the H_2O_2 byproduct was noticed—a steady-state enzymatic turnover rate upon introducing glucose, which is straightly associated with glucose concentration in solution. The inhibition of the steady-state enzymatic reaction of glucose oxidase thru a quantitative reduction in the steady-state rate, successively producing a metal ion biosensor with ultra-high sensitivity for heavy metals, for example, Ag^+ ions, through enzymatic inhibition. At 0.05 V of low-operating potential vs Hg/Hg_2SO_4, the developed biosensor of Ag^+ displayed 0.19 ± 0.04 ppb of detection limit ($\sigma=3$) in 20–200 nM of linear range, a sensitivity of $2.00\times10^8\pm0.06$ M^{-1}, sample recovery at $101\%\pm2\%$. With sensitivities of $2.69\times10^3\pm0.07$ and $1.45\times10^6\pm0.05$ M^{-1}, the Cu^{2+} and Co^{2+} heavy metal ions, correspondingly were also sensed using enzyme biosensor but attained to be considered a reduced amount of inhibition property. Using Cornish-Bowden and Dixon plots, the inhibition mode of GO_x was observed for Cu^{2+} and Ag^+ ions. A strong correlation was detected between the biosensor sensitivity and the inhibition constants [20].

Similarly, an effective and facile inkjet printing method has been utilized as a novel sort of ionic liquid (IL)–CNT–graphene film (GF), which is sandwiched structured. The IL–CNT–GF material determines excellent possessions comprising rapid charge transfer, large surface area, and enough surface-active sites. In the nanohybrid material, the synergistic effects of diverse components are present. High sensitivity, good selectivity, and a virtuous sensing performance with a low sensing limit of 0.2 nM for Pb^{2+} ions, and the low sensing limit was found to be 0.1 nM for Cd^{2+} with an extensive linear range up to 1 µM with S/N=3, which shown by the electrochemical determination. These developed electrochemical

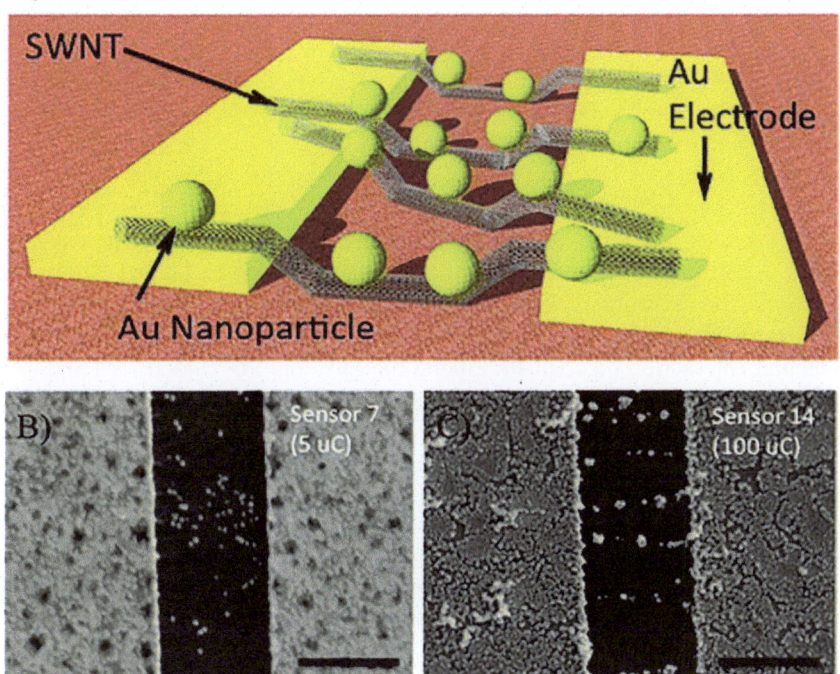

FIGURE 10.1 Development of CNTs through incorporation of AuNPs. (Reprinted with permission from Ref. [15]. Copyright 2011 American Chemical Society.)

sensing capacities offer it to be used [21]. Figure 10.1 shows the development of gold-incorporated single-walled carbon nanotubes (SWCNTs).

Electrochemical sensors using stripping voltammetry-based detection systems have been utilized for trace recognition of toxic metal ions in biologically applicable matrices endure an important challenge. Therefore, MWCNTs wrapped with chelating polymer have been developed as an alternative method to detect Cu^{2+} ions at the electrode surface selectively. A MWCNTs wrapped with poly-4-vinyl pyridine (P4VP) functionalized gold electrode modified with PET ($r = 1.5$ mm) has been used as a sensor. The detection through cyclic voltammetry (CV) beside the electrode surface for examination, a strong binding with Cu^{2+} ions done by the P4VP. With a 1.1–13.8 ppm of linear range (16.6–216 μM), and 4.9% of a RSD, the sensor showed a detection limit of 0.5 ppm, which is lower than the detection limit of 1.3 ppm according to EPA. Showing the accessibility of the developed detection method, estimation in lake water, tap water, deionized water, and ocean water showed similar outcomes. For continuous monitoring of natural waters, the development and design of heavy metal sensors represents an effective platform for participating in electrochemical determination applications analogous to polymeric chelation [22].

Improved differential pulse anodic stripping voltammetric method was utilized to develop the MWCNTs covalently functionalized with novel thiacalixarene (TCA) to detect Pb^{2+} ions in water samples. The GCE was modified with a TCA-MWCNT sensing probe used to select the metals on its surface. For electrochemical sensing of Pb^{2+} ions, the developed electrode material shows high sensitivity and outstanding selectivity through the excellent electronic properties of MWCNTs and the TCA combination. The detection limit was 4×10^{-11} mol L^{-1} for Pb^{2+} ions in 2×10^{-10} to 1×10^{-8} mol L^{-1} of concentration range, the stripping showed an extremely linear response with a regression coefficient value $R = 0.999$. Moreover, in the presence of an identical quantity of interfering Sn^{2+} metal ions, the detection of Pb^{2+} ions with concentration of 10^{-7} mol L^{-1} produced well-separated signals. Theoretical computations were accomplished to show the molecular interaction mechanism among the Sn^{2+} and Pb^{2+} metal ions and TCA molecules. The outcomes showed that because of the important electron delocalization among S atoms of the TCA molecule, the Sn^{2+} and Pb^{2+} ions could steadily adsorb on the surface of TCA molecules [23]. The sensing of Eu^{3+} metal ions is demonstrated in Figure 10.2 using CNTs.

A fast, sensitive, and selective field-effect transistor (FET) sensor, that is, SWCNTs, was developed to detect Hg^{2+} ions in a water sample. In mixed metal ion solution, the developed sensor offers selectivity toward Hg^{2+} ions. Through a strong redox reaction among Hg^{2+} ions and SWCNTs, the detection process depends on the abnormal response of conductance of SWCNT to the exposure of Hg^{2+} ions. The developed sensor showed 10 nM of LOD toward Hg^{2+} ions in water. For detecting Hg^{2+} ions, the sensor also showed a steep slope in a detection range from 10 nM to 1 mM with ultra-high sensitivity [24].

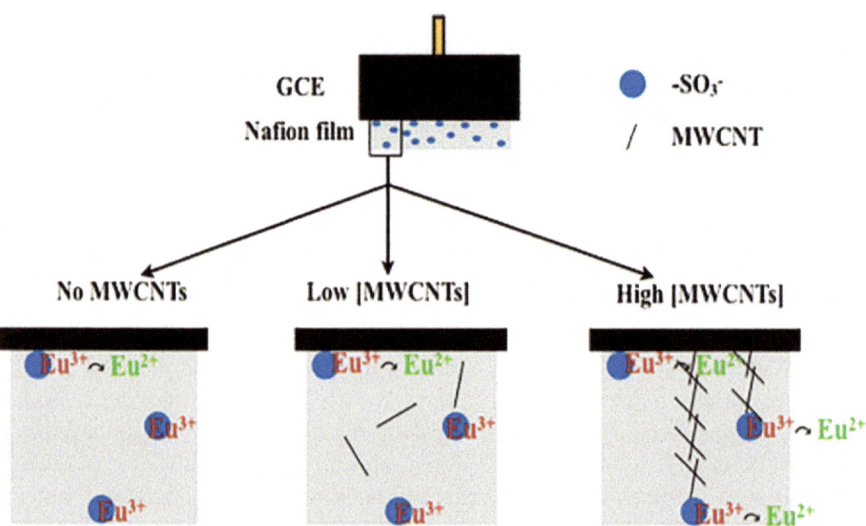

FIGURE 10.2 Different CNTs for the detection of heavy metal ions. (Reprinted with permission from Ref. [25]. Copyright 2014 American Chemical Society.)

In continuation, to sense trace concentrations of europium (Eu^{3+}) metal ions, a glassy carbon (GC) electrode was developed using MWCNTs loaded novel Nafion film. The dispersion of MWCNTs in Nafion allowed through interaction among the hydrophobic backbone of Nafion and the sidewalls of MWCNTs, which could be coated on the surface of the GC electrode as a thin film. By reducing the diffusion distance of Eu^{3+} to the electrode surface, the electrochemical response to Eu^{3+} proficiently expanded the electrode surface into the Nafion film. It was found to be ~10 times enhanced thru MWCNT concentrations among 0.5 and 2 mg mL^{-1}. In contrast to bare Nafion, no substantial enhancement was attained in signal at low concentrations of MWCNT up to 0.25 and 0.5 mg mL^{-1}. The SEM and electrochemical impedance spectroscopy (EIS) were utilized to characterize the MWCNT–Nafion film structure. This was followed by electrochemical characterization of Eu^{3+} ion using preconcentration voltammetry and CV. A linear range of 1–100 nM was experimented with for Eu^{3+} ion determination through osteryoung square wave voltammetry, which showed 0.37 nM of LOD with S/N=3 under the optimized conditions after a 480 s of preconcentration time [25].

Similarly, AuNPs functionalized SWCNT has been developed to detect Cu^{2+} and Pb^{2+} ions with ultra-high sensitivity. Using a CV method, the Au NPs were deposited on CNT film through an electrochemical method. The developed film could be measured through the gold precursor concentration and potential scanning cycle, which displayed a homogeneous density and size. Under optimized conditions, the detection of Cu^{2+} and Pb^{2+} ions was done through the square wave stripping voltammetry (SWSV) simultaneously. In a mixture of Cu^{2+} and Pb^{2+} solution, the AuNP–SWCNT electrode showed detection limit of 0.613 ppb ($R^2=0.991$) and 0.546 ppb ($R^2=0.984$) for Cu^{2+} and Pb^{2+} ions, correspondingly in the concentration range of 22.29–3.31 ppb with a good linear response. The probe also showed a high upsurge sensitivity with S/N=3 and $n=5$. In repetitive measurements for Cu^{2+} and Pb^{2+} ions, the developed electrode showed height reproducibility with 2.6% and 4.2% of relative standard deviation, correspondingly. The selectivity of an experiment performed in the presence of other metal ions, that is, As^{3+}, Sb^{3+}, Na^+, Ca^{2+}, and Zn^{2+} ions, does not significantly affect the selectivity. The synthesized technique is an alternative to the reliable and rapid sensing of toxic metals in environmental samples [26]. Various CNT-based electrodes are comprised in Table 10.1 to detect toxic metals.

10.3   ELECTRODES MODIFIED WITH CNT

The electrode materials have been developed as CNTs changed with N,N′,N″,N‴-tetrasalicylidene-3,3′-diaminobenzidine (H4tsdb) for the detection of Hg^{2+} ions using SWASV. The N,N′,N″,N‴-tetrasalicylidene-3,3′-diaminobenzidine (Octadentate ligand) was used to develop a detection probe by condensation of 3″3-diaminobenzidine and salicylaldehyde. Various analytical tools have been used for the characterization of the ligand, such as UV–vis spectroscopy, FTIR, and Proton-Nuclear magnetic resonance (H^1-NMR) spectroscopy. The SEM and EIS were used to characterize the developed electrode's surface morphology.

TABLE 10.1
Carbon Nanotube (CNT) Based Electrode Sensors for Heavy Metal Ion Detection

Electrode	Covalent Modification	Target Ion	Simultaneous Detection?	Real Sample	Linear Range	LOD	References
MWCNTs/Nafion/PDMcT/Bi-GC	-SH	Pb, Cd	Yes	Tap and spring water	0.1–22, 0.05–20 ppb	0.03, 0.05 ppb	[27]
MWCNT	-GC -NH$_2$	Cd, Zn, Cu, Hg	Yes	Reservoir water	-	0.003, 0.021, 0.014, 0.029 ppb	[28]
MWCNTs-GC	Polyhistidine	Cu	-	Tap and groundwater	0.032–7.9 ppm	4.8 ppb	[29]
MWCNTs/Bi-GC	-SH, -NH$_2$, -COO-, -OH	Pb, Cd	Yes	-	2–50 ppb	0.3, 0.4 ppb	[30]
(SWCNTs/Hg/Bi)-GC	-COO-, -OH	Pb, Cd, Zn	Yes	River water	0.005–1.1, 0.5–11, 10–130 ppb	0.12, 0.076, 0.23 ppb	[31]
SWCNTs-AuNP-GC	L–Cystein	Cu	-	River water	0.0060–8.9 ppb	0.0012 ppb	[32]
CNTs/Nafion-GC	-NH$_2$, -COO-, -OH	Pb, Cd	Yes	Tap and groundwater	0.0021–14.5 ppm 0.011–11.24 ppm	0.21, 0.56 ppb	[33]
MWCNTs/5-Br-PADAP/GC	-COO-, -OH	Pb	-	Spring, drinking, wastewater	0.9–114 ppb	0.1 ppb	[34]
MWCNTs/CTS/GA-SPE	-COO-, -OH	Pb	-	Natural water	0.099–2 ppb	11.8 ppb	[35]
MWCNTs/Bi/Nafion-SPE	-COO-, -OH	Pb, Cd, Zn	Yes	-	-	0.7, 1.5, and 11.1 nmol L^{-1}	[36]

To detect Hg^{2+} in $NaNO_3$ of 0.1 M concentration at pH 5, experimental parameters were optimized, for example, pH, supporting electrolytes, and preconcentration effect. For Hg^{2+} ions sensing, the electrochemical sensor showed the detection limit of 0.8 nM and a 2.4–220 nM of wide linear range under the optimized condition with (S/N ≤ 3). The five consecutive determinations showed 3.0% of the RSD. On the Hg^{2+} ion electrochemical response, the interference studies showed an insignificant effect from mixed metal ion solution, for example, Cd^{2+}, Mn^{2+}, Ni^{2+}, Pb^{2+}, Zn^{2+}, and Cu^{2+} ions. To recognize Hg^{2+} ions in real water samples, the developed electrode showed an outstanding capacity for selectivity, reproducibility, and fruitful application [37].

In an electrochemical cell, using all three electrodes, a sensor has been developed, which is based on CNT thread. A micro quasi-reference electrode was synthesized to define the microelectrode area. In this, the CNT thread is partially insulated with a thin polystyrene coating and used as the working electrode, for this, the bare CNT thread works as an auxiliary electrode, and CNT thread is incorporated with Ag thru electroplating and to form a layer of AgCl, it is anodizing in chloride solution. The electrode of CNT thread coated with Ag|AgCl offered a stable potential contrary to the Ag|AgCl reference electrode of conventional liquid-junction type. The auxiliary electrode of the CNT thread provides a stable current, which is like an auxiliary electrode of Pt wire. Thru the ASV, all the three CNT thread electrode cells used for the real-time sensing of toxicants have been assessed as a microsensor. The developed sensor probe has been utilized for Pb^{2+}, Cu^{2+}, and Hg^{2+} ions. The calculated LOD was 0.57, 0.53, and 1.05 nM, correspondingly for Hg^{2+}, Cu^{2+}, and Pb^{2+} ions at a deposition time of 120 s. The dimensions decreased to the microscale of the three conventional electrochemical cell electrodes using these electrodes [38].

In a similar study, a stainless steel electrode immobilized with urease was utilized to detect toxic metals in water samples through the facile synthesis of polypyrrole (PPy) deposited MWCNT composites. In 0.1 M aqueous solution of dodecylbenzene sulfonic acid, the nanocomposites were electrochemically developed on stainless steel using functionalized MWCNT and PPy. Functionalized MWCNT of diverse weight% was utilized relating to pyrrole to synthesize nanocomposite films. For the immobilization of urease, the films were used deposited on the electrode. The cross-linking urease (2 mg mL^{-1}) was used to immobilize at 0.1 M phosphate buffer with 7.0 pH through 0.1% glutaraldehyde. Diverse analytical tools were used, that is, CV, FTIR, and SEM, to characterize the developed nanocomposites. To sense various concentrations of heavy metals, the urease biosensors have been used after standardization. In 1–10 mM of the concentration range, the synthesized probe exhibited the lowest detection limit for Pb^{2+} and Cd^{2+} metal ions [39].

Similarly, for electrochemical sensing of Pb^{2+} ions, the selectivity improvement of GO–MWCNTs–L-cysteine (ErGO–MWNTs–L-cys) reduced electrochemically and modified GCE nanocomposite developed using a drop-casting method. Primarily, a cost-effective and facile technique has been utilized at room temperature to synthesize the nanocomposite of GO–MWNTs–L-cys. To inhibit the hydrophobicity of MWNTs, the as-synthesized GO–MWNTs–L-cys showed various stable aqueous dispersions because of the high hydrophilic nature of GO components. Formerly, the electrochemical conductivity of GCE (ErGO–MWNTs–L-cys/GCE) modified with ErGO–MWNTs–L-cys nanocomposite was enhanced thru the direct electrochemical reduction of GO–MWNTs–L-cys nanocomposite. The analytical tools such as UV–vis spectroscopy, XRD, attenuated total reflection infrared (ATR-IR), AFM, and Raman spectroscopy were used to characterize the individual components of the probe and GO–MWNTs–L-cys nanocomposites. The synergistic effect of ErGO–MWNTs–L-cys nanocomposite was confirmed using CV measurements and EIS in $[Fe(CN)_6]^{3-/4-}$ redox solution. The electrochemical reduction degrees, accumulation time, and pH are the optimizing parameters for the development of sensor electrodes. The Pb^{2+} ions were examined for electrochemical competence of modified electrodes, and under optimal conditions, it showed considerable enhancement at the ErGO–MWNTs–L-cys/GCE. Differential pulse anodic stripping voltammetry (DPASV) technique was utilized to detect Pb^{2+} ions using ErGO–MWNTs–L-cys/GCE nanocomposite. With the detection limit of $0.1\ \mu g\ L^{-1}$ with S/N = 3, the calibration plots showed 0.2–$40\ \mu g\ L^{-1}$ of concentration range for the linear relationship among Pb^{2+} ions and anodic current. The results showed that the ErGO–MWNTs–L-cys/GCE showed 2.15% of RSD with exhibited stable results and satisfied selectivity [40].

Surfaces of CNTs electrode changed chemically used to detect toxic metals using a sensitive voltammetric technique. Cysteine is used to functionalize CNTs covalently after casting on electrode surfaces. For the sensing of metals, cysteine shows high affinities, which are amino acid. Here, the electrode surfaces of CNT changed with cysteine accumulate the toxic metals, which was confirmed by the analysis of differential pulse anodic stripping voltammetry. The technique was improved regarding reduction potential, reduction time, and accumulation time. For Cu^{2+} and Pb^{2+} heavy metal ions, the detection limits were 15 and 1 ppb, respectively. They did the detection process in spiked lake water for Pb^{2+} and Cu^{2+} ions through the developed method. The average recoveries of Pb^{2+} and Cu^{2+} ions with relative standard deviations of 7.53% and 8.43% were found at 94.5% and 96.2%. Through the modified CNTs, the potential was also established for simultaneous sensing of toxic ions [41].

Toxic ions own a grave threat to animals and humans; therefore, it is vital to detect food and the environment. A simultaneous electrochemical sensor, such as Fe_3O_4/MWCNTs, Fe_3O_4/F-MWCNTs, and Fe_3O_4 NPs-based nanocomposites, has been developed using a hydrothermal method to detect heavy metal ions. The catalytic performances of the Fe_3O_4/F-MWCNTs sensor, in contrast to the three electrochemical sensors, showed greater performance toward sensing Pb^{2+}, Cd^{2+}, Hg^{2+}, and Cu^{2+} ions. The Pb^{2+}, Cd^{2+}, Hg^{2+}, and Cu^{2+} metal ions could be detected using the developed probe, which showed the sensitivity of 125.91, 108.79, 312.65, and 160.85 $\mu A\ mM^{-1}\,cm^{-2}$, correspondingly, which was noticeably advanced contrary to Fe_3O_4 and Fe_3O_4/MWCNTs. For Pb^{2+}, Cd^{2+}, Hg^{2+}, and Cu^{2+} ions, a wider linear detection range of 0.5–30.0, 0.5–30.0, 0.5–20.0, and 0.5–30.0 μM shown by the Fe_3O_4/F-MWCNTs sensor, correspondingly. For the detection of Pb^{2+}, Cd^{2+}, Hg^{2+}, and Cu^{2+}, the developed Fe_3O_4/F-MWCNTs sensor showed 0.08, 0.05, 0.05, and 0.02 nM of limit of detections with (S/N ratio = 3) correspondingly. In soybean and river water samples, the typical methods, that is, ICPMS, atomic fluorescence spectrometry (AFS), the low-cost Fe_3O_4/F-MWCNTs sensor, showed outstanding potential. In recovery, selectivity, reproducibility, and stability, the as-synthesized sensor showed outstanding performances. The developed sensor offers a promising method to display multiple targets in the food and environment [42].

Low-site density CNTs based nanoelectrode arrays (NEAs) have been used to detect toxic metals using an ultrasensitive voltammetric technique. Corresponding to a low background current, the NEAs were synthesized thru sealing with a passive epoxy layer of the sidewalls of CNTs that eliminates the electrode capacitance and decreases the current leakage, which offers a high S/N ratio. For voltammetric detection of trace Pb^{2+} and Cd^{2+} ions at their sub-ppb level, the CNTs-NEAs have been developed coated with a bismuth film. The developed probe under optimum experimental conditions showed 0.04 mg L^{-1} of LOD. The novel carbon NEA sensing platform's fascinating performance grasps great potential [43].

The sensing in aqueous media of Pb^{2+} and Cd^{2+} ions, the PyTS–CNTs/Nafion/PGE sensor electrode have been developed using the DPASV technique. The developed sensor probe detected Pb^{2+} and Cd^{2+} ions with the detection at 0.02 and 0.8 µg L^{-1}, correspondingly [44].

The detection of toxic metals based on the SPR technique has been done using multi-walled CNT modified with polypyrrole on a gold layer composite. The novel sensor was developed for the toxic metals, for example, Fe^{2+}, Pb^{2+}, and Hg^{2+} ions. A polypyrrole layer and a multi-walled CNT functionalized with polypyrrole composite were compared for heavy metal ions' sensitivity. Because of the binding of Hg^{2+}, Pb^{2+}, and Fe^{2+} ions on the detection layer, the sensor's accuracy and sensitivity improved by using polypyrrole MWCNTs toward sensing in an aqueous solution. The sensor probe strongly bonded with Hg^{2+} ions in contrast to Fe^{2+} and Pb^{2+} ions. Near 0.3°–0.6°, an angle shift was produced due to the limitation of the sensor with 0.1 ppm of LOD [45].

For the detection of heavy metal, electrochemical sensors should have exclusive mechanical, chemical, and electronic properties in which CNTs make them extremely attractive materials [46]. In contrast to bulk carbon electrodes, detection using CNTs holds the benefits of their bulk properties, for example, the fast electron transfer rate and increased electrode surface area. In contrast to the traditional macroelectrodes, the nanomodified electrodes have an improved rate of mass transport, big mass sensitivity, a higher S/N ratio, and a reduced effect of the resistance of solution [47]. For the sensing of Hg^{2+} and Cd^{2+} metal ions in wastewater, the MWCNTs/peptide sensor probe has been developed. The MWCNTs/peptide sensor was synthesized using the CV technique and showed the lowest detection limit of 9.068×10^{-10} and 2.749×10^{-8} M [48]. They observed a linear relationship of concentration from 0.5 to 8 mg L^{-1} among the cadmium concentration and the stripping current.

Some other MWCNTs have been developed and bismuth films to detect toxic metal ions [49–51]. The mixture of MWCNTs with screen-printed electrodes (SPE) and mercury nanodroplets has been examined toward real-time detection of Cu^{2+}, Pb^{2+}, and Cd^{2+} ions [52]. To evade extra environmental problems, the usage of free-mercury electrodes should be of greater interest, though the concentration of mercury is low used in the study. The metal nanotubes are a virtuous substitute for metal CNTs and NPs. Xu et al. reported the oxidative detection of As^{3+} by linear scan voltammetry on exceedingly ordered electrodes of platinum nanotube array [53]. The developed technique has practical utility for ecological monitoring of As^{3+} ions considering the reproducibility and detection limit.

Electrode modification uses the mixture of MNPs with CNTs [54] to sense As^{3+} ions at their trace level. The electroactive area ensuing from smaller size Pt NPs was improved, with the upsurge of the sizeable local mass transport rates. The electrode developed by the use of CNTs corresponds toward detection of As^{3+} ions for limits of detection up to ppb to ppt level.

10.4 CONCLUSION

Carbon materials incorporated with other detection methods improve detection sensitivity and optimize detection, owing to their highly stable chemical properties and ultra-high conductivity. Carbon-modified electrodes have straightforward applications in situ, online, and real-time analyses and show excellent application prospects, especially in the electrochemical testing method. The development of new composite electrodes and continued advancements fascinated the researcher's attention through carbon materials to progress the ant interference ability and selectivity of electrochemical detection technology. These electrodes expand the application scope of carbon-based composite electrodes and increase the research application in unconventional environments and the service life of electrodes.

REFERENCES

1. Gooding, J. Justin, Rahmat Wibowo, Jingquan Liu, Wenrong Yang, Dusan Losic, Shannon Orbons, Freya J. Mearns, Joe G. Shapter, and D. Brynn Hibbert. "Protein electrochemistry using aligned carbon nanotube arrays." *Journal of the American Chemical Society* 125, no. 30 (2003): 9006–9007. doi:10.1021/ja035722f.
2. Peigney, Alain, Ch Laurent, Emmanuel Flahaut, R. R. Bacsa, and Abel Rousset. "Specific surface area of carbon nanotubes and bundles of carbon nanotubes." *Carbon* 39, no. 4 (2001): 507–514. doi:10.1016/S0008-6223(00)00155-X.
3. Wang, Tingting, Hemanthi D. Manamperi, Wei Yue, Bill L. Riehl, Bonnie D. Riehl, Jay M. Johnson, and William R. Heineman. "Electrochemical studies of catalyst free carbon nanotube electrodes." *Electroanalysis* 25, no. 4 (2013): 983–990. doi:10.1002/elan.201200458.
4. Zhao, Ke, Andrei Veksha, Liya Ge, and Grzegorz Lisak. "Near real-time analysis of para-cresol in wastewater with a laccase-carbon nanotube-based biosensor." *Chemosphere* 269 (2021): 128699. doi:10.1016/j.chemosphere.2020.128699.
5. Wang, Tingting, Daoli Zhao, Noe Alvarez, Vesselin N. Shanov, and William R. Heineman. "Optically transparent carbon nanotube film electrode for thin layer spectroelectrochemistry." *Analytical Chemistry* 87, no. 19 (2015): 9687–9695. doi:10.1021/acs.analchem.5b01784.
6. Wang, Joseph, Randhir P. Deo, Philippe Poulin, and Maryse Mangey. "Carbon nanotube fiber microelectrodes." *Journal of the American Chemical Society* 125, no. 48 (2003): 14706–14707. doi:10.1021/ja037737j.

7. Nugent, J. M., K. S. V. Santhanam, A. A. Rubio, and P. M. Ajayan. "Fast electron transfer kinetics on multiwalled carbon nanotube microbundle electrodes." *Nano Letters* 1, no. 2 (2001): 87–91. doi:10.1021/nl005521z.

8. Liu, Guodong, Yuehe Lin, Yi Tu, and Zhifeng Ren. "Ultrasensitive voltammetric detection of trace heavy metal ions using carbon nanotube nanoelectrode array." *Analyst* 130, no. 7 (2005): 1098–1101. doi:10.1039/B419447K.

9. Wang, Joseph, Guodong Liu, and M. Rasul Jan. "Ultrasensitive electrical biosensing of proteins and DNA: Carbon-nanotube derived amplification of the recognition and transduction events." *Journal of the American Chemical Society* 126, no. 10 (2004): 3010–3011. doi:10.1021/ja031723w.

10. Gong, Kuanping, Supriya Chakrabarti, and Liming Dai. "Electrochemistry at carbon nanotube electrodes: Is the nanotube tip more active than the sidewall?" *Angewandte Chemie*, 120, no. 29 (2008): 5526–5530. doi:10.1002/ange.200801744.

11. Viry, Lucie, Alain Derré, Patrick Garrigue, Neso Sojic, Philippe Poulin, and Alexander Kuhn. "Optimized carbon nanotube fiber microelectrodes as potential analytical tools." *Analytical and Bioanalytical Chemistry* 389, no. 2 (2007): 499–505. doi:10.1007/s00216-007-1467-9.

12. Jacobs, Christopher B., Ilia N. Ivanov, Michael D. Nguyen, Alexander G. Zestos, and B. Jill Venton. "High temporal resolution measurements of dopamine with carbon nanotube yarn microelectrodes." *Analytical Chemistry* 86, no. 12 (2014): 5721–5727. doi:10.1021/ac404050t.

13. Güell, Aleix G., Katherine E. Meadows, Petr V. Dudin, Neil Ebejer, Julie V. Macpherson, and Patrick R. Unwin. "Mapping nanoscale electrochemistry of individual single-walled carbon nanotubes." *Nano Letters* 14, no. 1 (2014): 220–224. doi:10.1021/nl403752e.

14. Zhao, Daoli, Xuefei Guo, Tingting Wang, Noe Alvarez, Vesselin N. Shanov, and William R. Heineman. "Simultaneous detection of heavy metals by anodic stripping voltammetry using carbon nanotube thread." *Electroanalysis* 26, no. 3 (2014): 488–496. doi:10.1002/elan.201300511.

15. McNicholas, Thomas P., Kang Zhao, Changheng Yang, Sandra C. Hernandez, Ashok Mulchandani, Nosang V. Myung, and Marc A. Deshusses. "Sensitive detection of elemental mercury vapor by gold-nanoparticle-decorated carbon nanotube sensors." *The Journal of Physical Chemistry C* 115, no. 28 (2011): 13927–13931. doi:10.1021/jp203662w.

16. Guo, Xuefei, Woo Hyoung Lee, Noe Alvarez, Vesselin N. Shanov, and William R. Heineman. "Detection of trace zinc by an electrochemical microsensor based on carbon nanotube threads." *Electroanalysis* 25, no. 7 (2013): 1599–1604. doi:10.1002/elan.201300074.

17. Zhao, Daoli, Tingting Wang, Daewoo Han, Cory Rusinek, Andrew J. Steckl, and William R. Heineman. "Electrospun carbon nanofiber modified electrodes for stripping voltammetry." *Analytical Chemistry* 87, no. 18 (2015): 9315–9321. doi:10.1021/acs.analchem.5b02017.

18. Yue, Wei, Bill L. Riehl, Nebojsa Pantelic, Kevin T. Schlueter, Jay M. Johnson, Robert A. Wilson, Xuefei Guo, Edward E. King, and William R. Heineman. "Anodic stripping voltammetry of heavy metals on a metal catalyst free carbon nanotube electrode." *Electroanalysis* 24, no. 5 (2012): 1039–1046. doi:10.1002/elan.201200065.

19. Hui, Xue, Md Sharifuzzaman, Sudeep Sharma, Xing Xuan, Shipeng Zhang, Seok Gyu Ko, Sang Hyuk Yoon, and Jae Yeong Park. "High-performance flexible electrochemical heavy metal sensor based on layer-by-layer assembly of Ti_3C_2T x/MWNTs nanocomposites for noninvasive detection of copper and zinc ions in human biofluids." *ACS Applied Materials & Interfaces* 12, no. 43 (2020): 48928–48937. doi:10.1021/acsami.0c12239.

20. Rust, Ian M., Jacob M. Goran, and Keith J. Stevenson. "Amperometric detection of aqueous silver ions by inhibition of glucose oxidase immobilized on nitrogen-doped carbon nanotube electrodes." *Analytical Chemistry* 87, no. 14 (2015): 7250–7257. doi:10.1021/acs.analchem.5b01224.

21. Dong, Shuang, Zhengyun Wang, Muhammad Asif, Haitao Wang, Yang Yu, Yulong Hu, Hongfang Liu, and Fei Xiao. "Inkjet printing synthesis of sandwiched structured ionic liquid-carbon nanotube-graphene film: Toward disposable electrode for sensitive heavy metal detection in environmental water samples." *Industrial & Engineering Chemistry Research* 56, no. 7 (2017): 1696–1703. doi:10.1021/acs.iecr.6b04251.

22. Fernando, PU Ashvin Iresh, Erik Alberts, Matthew W. Glasscott, Anton Netchaev, Jason D. Ray, Keith Conley, Rishi Patel et al. "In situ preconcentration and quantification of Cu^{2+} via chelating polymer-wrapped multiwalled carbon nanotubes." *ACS Omega* 6, no. 8 (2021): 5158–5165. doi:10.1021/acsomega.0c04776.

23. Wang, Li, Xiaoya Wang, Guosheng Shi, Cheng Peng, and Yihong Ding. "Thiacalixarene covalently functionalized multiwalled carbon nanotubes as chemically modified electrode material for detection of ultratrace Pb^{2+} ions." *Analytical Chemistry* 84, no. 24 (2012): 10560–10567. doi:10.1021/ac302747f.

24. Kim, Tae Hyun, Joohyung Lee, and Seunghun Hong. "Highly selective environmental nanosensors based on anomalous response of carbon nanotube conductance to mercury ions." *The Journal of Physical Chemistry C* 113, no. 45 (2009): 19393–19396. doi:10.1021/jp908902k.

25. Wang, Tingting, Daoli Zhao, Xuefei Guo, Jaime Correa, Bill L. Riehl, and William R. Heineman. "Carbon nanotube-loaded nafion film electrochemical sensor for metal ions: Europium." *Analytical Chemistry* 86, no. 9 (2014): 4354–4361. doi:10.1021/ac500163f.

26. Bui, Minh-Phuong Ngoc, Cheng Ai Li, Kwi Nam Han, Xuan-Hung Pham, and Gi Hun Seong. "Simultaneous detection of ultratrace lead and copper with gold nanoparticles patterned on carbon nanotube thin film." *Analyst* 137, no. 8 (2012): 1888–1894. doi:10.1039/C2AN16020J.

27. He, Xiuhui, Zhaohong Su, Qingji Xie, Chao Chen, Yingchun Fu, Li Chen, Ying Liu et al. "Differential pulse anodic stripping voltammetric determination of Cd and Pb at a bismuth glassy carbon electrode modified with Nafion, poly (2, 5-dimercapto-1, 3, 4-thiadiazole) and multiwalled carbon nanotubes." *Microchimica Acta* 173, no. 1–2 (2011): 95–102. doi:10.1007/s00604-010-0541-8.

28. Wei, Yan, Ran Yang, Xing Chen, Lun Wang, Jin-Huai Liu, and Xing-Jiu Huang. "A cation trap for anodic stripping voltammetry: NH_3–plasma treated carbon nanotubes for adsorption and detection of metal ions." *Analytica Chimica Acta* 755 (2012): 54–61. doi:10.1016/j.aca.2012.10.021.

29. Dalmasso, Pablo R., María L. Pedano, and Gustavo A. Rivas. "Electrochemical determination of Cu (II) using a glassy carbon electrode modified with multiwall carbon nanotubes dispersed in polyhistidine." *Electroanalysis* 27, no. 9 (2015): 2164–2170. doi:10.1002/elan.201500116.

30. Li, Xueying, Haihui Zhou, Chaopeng Fu, Fei Wang, Yujie Ding, and Yafei Kuang. "A novel design of engineered multi-walled carbon nanotubes material and its improved performance in simultaneous detection of Cd^{2+} and Pb^{2+} by square wave anodic stripping voltammetry." *Sensors and Actuators B: Chemical* 236 (2016): 144–152. doi:10.1016/j.snb.2016.05.149.

31. Ouyang, Ruizhuo, Zhenqian Zhu, Clarissa E. Tatum, James Q. Chambers, and Zi-Ling Xue. "Simultaneous stripping detection of Zn^{2+}, Cd^{2+} and Pb^{2+} using a bimetallic Hg–Bi/single-walled carbon nanotubes composite electrode." *Journal of Electroanalytical Chemistry* 656, no. 1–2 (2011): 78–84. doi:10.1016/j.jelechem.2011.01.006.

32. Fu, Xu-Cheng, Ju Wu, Jun Li, Cheng-Gen Xie, Yi-Shu Liu, Yu Zhong, and Jin-Huai Liu. "Electrochemical determination of trace copper (II) with enhanced sensitivity and selectivity by gold nanoparticle/single-wall carbon nanotube hybrids containing three-dimensional l-cysteine molecular adapters." *Sensors and Actuators B: Chemical* 182 (2013): 382–389. doi:10.1016/j.snb.2013.02.074.

33. Joshi, Anju, and Tharamani C. Nagaiah. "Nitrogen-doped carbon nanotubes for sensitive and selective determination of heavy metals." *RSC Advances* 5, no. 127 (2015): 105119–105127. doi:10.1039/C5RA15944J.

34. Salmanipour, Ashraf, and Mohammad Ali Taher. "An electrochemical sensor for stripping analysis of Pb (II) based on multiwalled carbon nanotube functionalized with 5-Br-PADAP." *Journal of Solid State Electrochemistry* 15, no. 11 (2011): 2695–2702. doi:10.1007/s10008-010-1197-3.

35. Vicentini, Fernando Campanhã, Tiago Almeida Silva, Allan Pellatieri, Bruno C. Janegitz, Orlando Fatibello-Filho, and Ronaldo Censi Faria. "Pb (II) determination in natural water using a carbon nanotubes paste electrode modified with crosslinked chitosan." *Microchemical Journal* 116 (2014): 191–196. doi:10.1016/j.microc.2014.05.008.

36. Mandil, Adil, Rasa Pauliukaite, Aziz Amine, and Christopher MA Brett. "Electrochemical characterization of and stripping voltammetry at screen printed electrodes modified with different brands of multiwall carbon nanotubes and bismuth films." *Analytical Letters* 45, no. 4 (2012): 395–407. doi:10.1080/00032719.2011.644742.

37. Gayathri, Jayagopi, Kumar Sangeetha Selvan, and Sangilimuthu Sriman Narayanan. "Fabrication of carbon nanotube and synthesized octadentate ligand modified electrode for determination of Hg (II) in Sea water and Lake water using square wave anodic stripping voltammetry." *Sensing and Bio-Sensing Research* 19 (2018): 1–6. doi:10.1016/j.sbsr.2018.02.006.

38. Zhao, Daoli, David Siebold, Noe T. Alvarez, Vesselin N. Shanov, and William R. Heineman. "Carbon nanotube thread electrochemical cell: Detection of heavy metals." *Analytical Chemistry* 89, no. 18 (2017): 9654–9663. doi:10.1021/acs.analchem.6b04724.

39. Meshram, B. H., S. B. Kondawar, A. P. Mahajan, R. P. Mahore, and D. K. Burghate. "Urease immobilized polypyrrole/multi-walled carbon nanotubes composite biosensor for heavy metal ions detection." *Journal of the Chinese Advanced Materials Society* 2, no. 4 (2014): 223–235. doi:10.1080/22243682.2014.935953.

40. Al-Gahouari, Theeazen, Gajanan Bodkhe, Pasha Sayyad, Nikesh Ingle, Manasi Mahadik, Sumedh M. Shirsat, Megha Deshmukh, Nadeem Musahwar, and Mehendra Shirsat. "Electrochemical sensor: L-Cysteine induced selectivity enhancement of electrochemically reduced graphene oxide–multiwalled carbon nanotubes hybrid for detection of lead (Pb^{2+}) ions." *Frontiers in Materials* 7 (2020): 68. doi:10.3389/fmats.2020.00068.

41. Morton, Jeffrey, Nathaniel Havens, Amos Mugweru, and Adam K. Wanekaya. "Detection of trace heavy metal ions using carbon nanotube-modified electrodes." *Electroanalysis: An International Journal Devoted to Fundamental and Practical Aspects of Electroanalysis* 21, no. 14 (2009): 1597–1603. doi:10.1002/elan.200904588.

42. Wu, Wenqin, Mingming Jia, Zhaowei Zhang, Xiaomei Chen, Qi Zhang, Wen Zhang, Peiwu Li, and Lin Chen. "Sensitive, selective and simultaneous electrochemical detection of multiple heavy metals in environment and food using a lowcost Fe_3O_4 nanoparticles/fluorinated multi-walled carbon nanotubes sensor." *Ecotoxicology and Environmental Safety* 175 (2019): 243–250. doi:10.1016/j.ecoenv.2019.03.037.

43. Liu, Guodong, Yuehe Lin, Yi Tu, and Zhifeng Ren. "Ultrasensitive voltammetric detection of trace heavy metal ions using carbon nanotube nanoelectrode array." *Analyst* 130, no. 7 (2005): 1098–1101. doi:10.1039/B419447K.

44. Jiang, Ruyuan, Niantao Liu, Sanshuang Gao, Xamxikamar Mamat, Yuhong Su, Thomas Wagberg, Yongtao Li, Xun Hu, and Guangzhi Hu. "A facile electrochemical sensor based on PyTS–CNTs for simultaneous determination of cadmium and lead ions." *Sensors* 18, no. 5 (2018): 1567. doi:10.3390/s18051567.

45. Sadrolhosseini, Amir Reza, A. S. M. Noor, Afarin Bahrami, H. N. Lim, Zainal Abidin Talib, and Mohd Adzir Mahdi. "Application of polypyrrole multi-walled carbon nanotube composite layer for detection of mercury, lead and iron ions using surface plasmon resonance technique." *PloS One* 9, no. 4 (2014): e93962. doi:10.1371/journal.pone.0093962.

46. Baughman, Ray H., Anvar A. Zakhidov, and Walt A. De Heer. "Carbon nanotubes – The route toward applications." *Science* 297, no. 5582 (2002): 787–792. doi:10.1126/science.1060928.

47. Merkoçi, Arben, Martin Pumera, Xavier Llopis, Briza Pérez, Manel Del Valle, and Salvador Alegret. "New materials for electrochemical sensing VI: Carbon nanotubes." *TrAC – Trends in Analytical Chemistry* 24, no. 9 (2005): 826–838. doi:10.1016/j.trac.2005.03.019.

48. Awizar, Denni Asra, Norinsan Kamil Othman, Azman Jalar, Abdul Razak Daud, I. Abdul Rahman, and N. H. Al-Hardan. "Nanosilicate extraction from rice husk ash as green corrosion inhibitor." *International Journal of Electrochemical Science* 8, no. 2 (2013): 1759–1769. doi:10.20964/2021.01.30.

49. Wang, Na, and Xiandui Dong. "Stripping voltammetric determination of Pb (II) and Cd (II) based on the multiwalled carbon nanotubes-Nafion-bismuth modified glassy carbon electrodes." *Analytical Letters* 41, no. 7 (2008): 1267–1278. doi:10.1080/00032710802052817.

50. Hwang, Gil Ho, Won Kyu Han, Joon Shik Park, and Sung Goon Kang. "Determination of trace metals by anodic stripping voltammetry using a bismuth-modified carbon nanotube electrode." *Talanta* 76, no. 2 (2008): 301–308. doi:10.1016/j.talanta.2008.02.039.

51. Xu, He, Liping Zeng, Sujie Xing, Yuezhong Xian, Guoyue Shi, and Litong Jin. "Ultrasensitive voltammetric detection of trace lead (II) and cadmium (II) using MWCNTs-nafion/bismuth composite electrodes." *Electroanalysis: An International Journal Devoted to Fundamental and Practical Aspects of Electroanalysis* 20, no. 24 (2008): 2655–2662. doi:10.1002/elan.200804367.

52. Song, Wei, Lei Zhang, Lei Shi, Da-Wei Li, Yang Li, and Yi-Tao Long. "Simultaneous determination of cadmium (II), lead (II) and copper (II) by using a screen-printed electrode modified with mercury nano-droplets." *Microchimica Acta* 169, no. 3–4 (2010): 321–326. doi:10.1007/s00604-010-0354-9.

53. Xu, He, LiPing Zeng, SuJie Xing, GuoYue Shi, Junshui Chen, YueZhong Xian, and Litong Jin. "Highly ordered platinum-nanotube arrays for oxidative determination of trace arsenic (III)." *Electrochemistry Communications* 10, no. 12 (2008): 1893–1896. doi:10.1016/j.elecom.2008.09.037.

54. Xu, He, Liping Zeng, Sujie Xing, Yuezhong Xian, and Litong Jin. "Microwave-irradiated synthesized platinum nanoparticles/carbon nanotubes for oxidative determination of trace arsenic (III)." *Electrochemistry Communications* 10, no. 4 (2008): 551–554. doi:10.1016/j.elecom.2008.01.028.

11 Nanofiber-Based Nanostructures as Plasmonic Nanosensors

11.1 INTRODUCTION

Increased pollution of environmental sources due to rapid urbanization and industrialization is a matter of concern worldwide [1–5]. Because of nondegradable and high tendency to ensue in the food chain and environmental systems, heavy metals seem to be contaminants of enormous biological and ecological concern. They cause grave complications for human health and the environment [6–8]. The heavy metal ion accumulation in the human body can severely damage the intestinal tract, skeletal system, central nervous system, mucous tissue, reproductive systems, kidneys, and liver [9]. Altogether, the permissible limits of these heavy metal ions set by the USEPA and WHO range from ppt to ppm [5]. The early sensing of contamination in the environment and living systems, the fabrication of inexpensive and rapid methods become highly required to detect heavy metals.

It can limit the application of various analytical methods thru inflexibility for on-site analysis, complex procedures, and expensive equipment. However, these methods, such as AAS, ICP–MS, XPS, ICP–OES, and MS, are very useful to offer precise quantitative information on metal ions [10]. Considering these limitations, the fabrication of a detection probe is a prime focus that gives selective on-site or in situ screening of metals, short response times, and high sensitivity. To attain these necessities, potential candidates comprise optical techniques [11], electrochemical [12], and laser-induced breakdown spectroscopy (LIBS) [13]. Optical sensors have established excessive consideration for the biosensing of diverse contaminant species because they allow real-time and on-site sensing deprived of the usage of intricate instrumentation of different spectroscopy [14–17]. The fruitful fabrication of optical and chemical sensors that are robust and sensible and essentially correlated with their platform type has been developed [18,19]. Novel methods are being developed to fabricate innovative optical biosensor platforms with current developments in nanomaterials. In terms of selectivity, sensitivity, portability, and multiplexed sensing competence, these materials may correspond to meaningful growth in the capacity of sensors [3,4]. Many nanomaterials have been effectively exploited to fabricate sensors depending on diverse optical signal transduction systems, for example, quantum dots, GO and graphene, magnetic and metallic nanomaterials, and CNTs [15]. In developing optical sensors with many applications, nanofibers can be applied through electrospinning in contrast

DOI: 10.1201/9781003128281-11

to other nanomaterials, which comprise easy detection, low cost, easy functionalization, and fabrication, and modified properties, such as structure, chemical composition, diameter, morphology, and porosity [20–22]. The electrospun nanofibers (ESNFs) optical sensors based on ESNFs are obtained as membranes/mats and can be easily detachable after the detection process. Therefore, these are recyclable and cost-effective [20].

The three main techniques are comprised here by which the optically active ESNFs can be attained:

- Through suitable chemical functionalization, the binding of optically active nanosystems to the polymer surface
- Using conjugated polymers, which may essentially emit/absorb light, and through inserting luminescent or absorbing chromophores or NPs into optically inert and transparent polymers [17].

For heavy metal optical sensing using ESNFs, the utmost used techniques are based on the emission or absorption of light. Using an indicator, the colorimetric sensing or absorption is accomplished upon binding to the analyte through alteration in its color. The alteration can be visible as well as spectroscopically determined. The analyte concentration is determined by altering the fluorophore's emission properties in light-emission methods after being excited through a definite wavelength [23]. The detection of toxicants using fluorescent or colorimetric techniques, this chapter comprises the current advancement for the fabrication of optical sensors based on ESNF. The different optically active electrospun nanofiber-based methods for detecting Fe^{3+}, Hg^{2+}, Cu^{2+}, Ni^{2+}, Cr^{3+}, and Pb^{2+} ions are also comprised in this chapter. Lastly, for sensing metal ions, the chapter specifies directions and trends toward the current developments to fabricate ESNFs and focuses on the future perspectives of the developed methods in sensing.

11.2 DETECTION OF HEAVY METALS USING NANOFIBER-BASED OPTICAL SENSORS

11.2.1 Fluorescence Technique for Optical Detection

The method of light emission thru a substance that has not been strongly heated is known as luminescence, where the emission is not related to thermal radiation. Luminescence is categorized into two classes, i.e., phosphorescence and fluorescence, depending on the emission rates (lifetime) and excited state. The main optical mechanisms used in optical sensors are resonance energy transfer (RET), stokes shift, and fluorescence quenching [24–27]. For diverse applications of sensing materials used in optical sensors, the high sensitivity is a significant feature offered thru fluorescence.

Here, the current research comprised here employed ESNFs for optical detection of diverse toxic metal ions using fluorescence. Recently, using the electrospinning method, the fabrication of fluorescent nanofibrous membrane (NFM) was developed by Ma and coworkers [28]. On hydrolyzed polyacrylonitrile (PAN) nanofiber membranes thru functionalization of dithioacetal-functionalized peryl-enediimide (DTPDI), the NFMs were attained. Because of the S–Hg^{2+} affinity, the DTPDI reaction with Hg^{2+} occurs when the Hg^{2+} aqueous solution is immersed in NFMs. Therefore, through π–π stacking, the C–S–C bonds of DTPDI was adsorbed on DTPDI and broken to produce hydrolysate. Numerous toxic metals such as Na^+, K^+, Cu^{2+}, Mg^{2+}, Zn^{2+}, Ca^{2+}, Fe^{3+}, Fe^{2+}, and Cr^{3+} ions have been used to select NFM. By an upsurge of intensity as a function of Hg^{2+} concentration, the NFMs showed sensitivity and high selectivity to detect Hg^{2+} ions with a 1.0×10^{-3} mg L^{-1} LOD. For diverse pollutants with no pertinent interfering, the study shows that the sensing of Hg^{2+} with NFMs was stable and reversible.

For heavy metal sensing, biomolecules, and pH, electrospun fiber membranes fascinated many multifunctional sensors. Therefore, Wang et al. [29] developed multifunctional ESNFs-based sensors using polymethylmethacrylate (PMMA) electrospun nanofibers (CPBQD/PMMA) incorporated with $CsPbBr_3$ perovskite quantum dots (CPBQDs). Because of the CPBQD/PMMA ESNFs high QY (88%), the sensing of pH, trypsin, and Cu^{2+} ions has been done using cyclam or Rhodamine 6G (R6G) and CPBQD/PMMA ESNFs, correspondingly. The FRET process is responsible for the efficient detection process. After attaching cyclam and the CPBQD/PMMA, the developed high-efficiency FRET process was used to probe Cu^{2+} ions. The PL quenching is used to measure the detection of Cu^{2+} using CPBQD/PMMA FM (fiber membrane)–cyclam because of the absorption band of cyclam–Cu^{2+} at 520 nm, which matches the emission band of the CPBQD/PMMA FM. For Cu^{2+} ions in a water sample, the developed probe showed 6.0×10^{-11} mg L^{-1} of the detection limit. In mixed metal ion solutions such as Mn^{2+}, Zn^{2+}, Fe^{2+}, K^+, Cr^{2+}, Na^+ and Co^{2+} ions, the developed CPBQD/PMMA–cyclam showed selectivity toward Cu^{2+} ions in the low-efficiency FRET processes between these ions and the CPBQD/PMMA. Figure 11.1 demonstrates the surface morphology and SPR images of the developed sensor before and after adding metal ions to the NP solution.

Similarly, Liang et al. [30], reported the detection of Hg^{2+} in aqueous solution using another multifunctional fluorescent ESNF attained thru the combination of magnetite NPs, the fluorescent probe poly (N-isopropylacrylamide)-co-(N-methylolacrylamide)-co-(acrylic acid), and 1-benzoyl-3-[2-(2-allyl-1,3-dioxo-2,3-dihydro-1Hbenzo[de]isoquinolin-6-amino) -ethyl]-thiourea (BNPTU). The developed NFMs showed high sensitivity toward mercury ions Hg^{2+}, temperature, and magnetism. Thru measurement of PL and absorbance, where a blue shift was observed of maximum emission, it showed 2.0×10^{-2} mg L^{-1} of the detection limit in an aqueous solution of Hg^{2+} ion. Compared to different metals, high sensitivity and selectivity for Hg^{2+} were detected. The same blue shift was detected between the 30°C and 60°C temperature range at fixed concentrations of Hg^{2+} ions.

Alternatively, confirming the selectivity for Hg^{2+} ion, an alteration in the PL and absorption peak was not detected for other metal ions. For applied and available advantages as multifunctional detection systems, the fluorescent magnetic ES nanofibers as naked-eye sensors can be utilized. Mesoporous ESNFs of

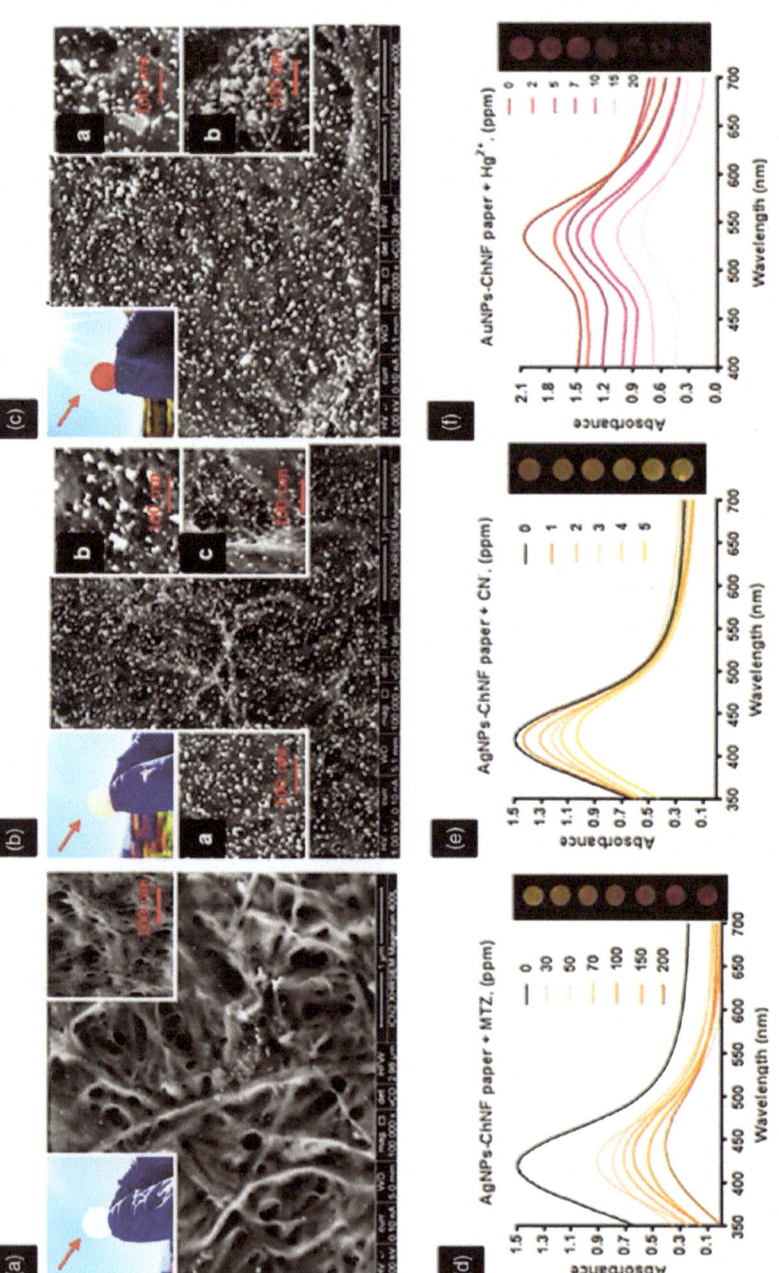

FIGURE 11.1 Optical detection platforms based on developed plasmonic ChNF paper. (A) Bare ChNF paper: digital photograph and SEM. (B) Developed AgNPs-ChNF paper: SEM (a) AgNPs-ChNF synthesized through in situ process; (b) AgNPs-ChNF with MTZ (100 ppm); (c) AgNPs-ChNF paper with CN⁻ (3 ppm). digital photograph of the fabricated AgNPs-ChNF paper (the inset image upper left). (C) AuNPs-ChNF paper: SEM (a) AuNPs-ChNF paper; (b) AuNPs-ChNF paper with Hg²⁺ (10 ppm). digital photograph of the fabricated AuNPs-ChNF paper (the inset image (left)). (D–F) Variations in UV–vis spectra and respective colors of (D) MTZ added AgNPs-ChNF paper (0–200 ppm), (E) CN⁻ added AgNPs-ChNF paper (0–5 ppm), and (F) Hg²⁺ added AuNPs-ChNF (0–20 ppm). (Reprinted with permission from Ref. [20]. Copyright 2020 American Chemical Society.)

silica/PAN (CD/mesoSiO$_2$/PAN) encapsulated with CDs have been demonstrated by Li et al. [31]. Because of their outstanding chemical and physical properties, the usage of CDs has been discovered in numerous fields. The researchers conserved the stability of the material, improving sensibility, and CDs' sensor capabilities, thru immobilization of the CDs into the polymer nanofiber. CD/mesoSiO$_2$/PANs are prominent PL sensors that are highly sensitive to Fe^{3+} ions. The CD/mesoSiO$_2$/PAN nanofibers exhibited improved selectivity, PL stability, and higher efficiency when related to the detection abilities of CDs and the developed probe. The CD/mesoSiO$_2$/PAN nanofibers and CDs in solution also showed selectivity for the detection of Fe^{3+}. Due to the interaction of phenol hydroxyl groups of CDs and Fe^{3+} ions, the developed probe showed selectivity further confirmed by the PL quenching signal. The designed probe showed the LOD as 0.2 mg L^{-1} for Fe^{3+} ions in the water sample. Their incorporation enhanced the luminescence properties of CDs on the electrospun nanofibers, resulting in the CD/mesoSiO$_2$/PAN nanofibers as utmost 'label-free 'and turn-off' fluorescence detection probe for HMs. The larger surface area of ESNFs in contrast to the thin films is the leading cause of their improved sensitivity.

Wu and coworkers [32] confirmed this enhancement by reporting the study where taking diverse molar ratios of PNNR1, PNNR2, and PNNR3 monomers (NIPAAm, NMA, and RHPMA), a comparison was established among the dip-coating films of poly[(N-isopropylacrylamide)-co-(N-hydroxymethyl acrylamide)-co-(4-rhodamine hydrazonomethyl-3-hydroxy-phenyl methacrylate)] [poly(NIPAAm-co-NMA-co-RHPMA)-PNNR] copolymers versus properties of electrospun nanofibers. The measurements of fluorescence confirmed that the [poly(NIPAAm-co-NMA-co-RHPMA)-PNNR] copolymers displayed a susceptible and selective detection of Cu^{2+} ions. The increased concentration of Cu^{2+} ions corresponds to a PL quenching process because of the paramagnetic effect of the d^9 system of the Cu^{2+} ion. A photoinduced electron transfer is suggested because of this process, from the excited state of singlet fluorophore (rhodamine) to paramagnetic metal center [32]. The developed PNNR nanofibers showed the lowest LOD for Cu^{2+} ions at 6.4×10^{-3} mg L^{-1}. The results showed that the PNNR nanofibers partook in potential water purification and pollutant separation and are used as fluorescent sensors of "on-off" types toward Cu^{2+} ions.

11.2.2 Colorimetry Technique for Optical Detection

The colorimetric sensing of a precise toxicant persuades alterations in the absorption band of the material. It results in color alterations that can be easily observed through the naked eye, as reported earlier. Subsequently, by simplifying the sensing apparatus, the sensing response measurement does not cause expensive and complex equipment [33,34]. However, the alteration in color may be exceedingly influenced through experimental parameters, such as visualization angle and luminosity. Next, the measurement of absorption using a UV–vis spectrometer, a color change, occurs to eliminate incorrect validation in colorimetric devices [35]. For more accurate systems consistent with the concentration and type of the pollutant, the sensor response is determined, and the absorbance band shift may be exactly examined. For the detection of heavy metals, organic dyes have been primarily developed with aromatic conjugated systems [36,37]. Sensing organic molecules through a complex process includes metal ion reprogenetics.

Raj et al. [38] detected Pb^{2+} ions in aqueous systems using curcumin, which is the utmost organic dye employed as a selective, biocompatible, and simple system. The cellulose acetate nanofibers (CC–CA) loaded with curcumin were developed through the strips fabricated by electrospinning of curcumin and cellulose acetate composite. For optimization of the operational conditions, a colorimetric study was utilized at diverse pH levels. The results showed that the absorbance intensity altered deprived of visual color change below pH 4. Deprived of important color or absorbance variations, better conditions were attained from pH 5 to 11 out of them, pH 5 was the optimum pH. The developed probe was tested for sensitivity ranging from 2.07×10^{-3} to 207 mg L^{-1} toward Pb^{2+} ions. The results showed that the sensor showed 4.1 mg L^{-1} of detection limit with a yellow to orange visual color change and a linear decrease of the absorbance upon adding analyte concentration to the probe solution. Through chelation of curcumin with Pb^{2+} complexation, the alteration in the color of probe solution occurs. In diverse metal ion solutions such as Ca^{2+}, Ba^{2+}, Cd^{2+}, Co^{2+}, Mg^{2+}, Cu^{2+}, Zn^{2+} and Ni^{2+} ions with 1 mM of concentration, the CC–CA strips showed as highly selective sensing probes for Pb^{2+}. For different ions, there are no color changes observed on the CC–CA strips. Through the α,β-unsaturated β-diketo moiety, a stable and robust complex is formed among Pb^{2+} ions and curcumin, which handles the selectivity of CC–CA strips toward the analyte. The enolic protons are substituted thru the metal ions during the complexation, changing the materials' color through changing the electronic configuration.

In a similar study for the pH-induced detection of Fe^{2+} ion, Saithongdee et al. [39] reported the zein NFM strips changed with curcumin as a chemosensor with advanced performance. The developed chemosensor showed 0.4 mg L^{-1} of detection limit with a visual color change in the presence of Fe^{2+} ions. To detect toxic metals in water samples, such outcomes establish the flexibility of the sensor's selectivity thru prudently selecting the allied material with the same optically active species. Dimethylglyoxime (DMG) acts thru complexing nickel and palladium ions, which is another organic aromatic molecule. In glass slides, the DMG combined polycaprolactam nanofibers (N_6-DMG@glass) were developed [40] for Ni^{2+} ions detection colorimetrically in laboratory wastewater and tap water. In the linear range from 5×10^{-3} to 1×10^{-1} mg L^{-1}, the developed strips showed a LOD at 2×10^{-3} mg L^{-1} with a visual color change from colorless to red. The selectivity test was carried out in mixed metal ion solutions, which exhibited substantial interference for Cu^{2+} in concentrations higher than 1 and 6×10^{-1} mg L^{-1}.

In contrast to other conjugated systems, such as organic dyes, the sensors have shown increased sensitivity with good capacity developed with conjugated polymers. In the long range, because of delocalized π-bonds, this behavior could be observed, which permits an increased sensitivity and a faster charge transfer. Changing the material's absorbance/reflectance and consequently the conjugation of the bonds thru a redox mechanism, the interaction among heavy metal ions and conducting polymers will occur in the sensing method.

In addition, PANI has imine and amine groups, and it is a classic conjugated polymer that can be made to suit and proton the amount of accessible H^+ ions in addition to the variation of its assets through the degree of oxidation, besides the change in different colors from white to blue. A second mechanism occurs through the protonation of imine and amine groups besides the oxidation process for developing sensors with PANI. The imine and amine groups will interact with the metal ions by changing the optical properties and the doping state. To change the optically active species into a fascinating form, Si et al. [41] developed the functionalized conjugated polymer membranes nanofiber. Therefore, the developed nanofiber membranes (PANI-LBNF) sensor exhibited the selective and sensitive interaction among Hg^{2+} ions and PANI. The nanofiber membranes were used to selectively detect Hg^{2+} ions in the water sample fabricated through the electrospinning method with an emeraldine base reduction to leucoemeraldine base hydrazine vapor. Consistent with the Hg^{2+} concentration, PANI-LBNF undergoes a compassionate and selective response either by an oxidation and protonation process, resulting in a drastic color alteration from white to blue, green, and yellow/green in the presence of Hg^{2+} ions. The intensity changes at 440 and 645 nm of the reflectance bands could characterize the sensor response based on the color alteration. They observed the lowest LOD of the developed probe as 1.0×10^{-3} mg L^{-1}, which can be detected with the naked eye.

Additionally, to confirm the color alteration of membrane accompanied by the Hg^{2+} concentration, an innovative colorimetric context was also utilized, which can be related to human insight, ensuring the consistency of usage of the PANI for this sensing method. Similarly, because of their optical properties based on size, metallic NPs are very interesting for colorimetric applications regarding inorganic materials [42]. Nanofiber membranes used as functionalizing agent on AuNPs substrate to develop optically active sensors [43]. For the colorimetric detection of Pb^{2+} ions, AuNPs (Au@GSH) incorporated with L- GSH used as a functionalizing agent on nylon 6/polyvinylidene-fluoride NFM surface. To stabilize the AuNPs, the GSH was added to a chelating agent, forming a covalent Au–S bond through the thiol group. Because of the well-dispersed Au@GSH functionalization, the Au@GSH functionalized NFM strips showed a vibrant pink color. Because of the decrease in distance between Au@GSH, the color of the strip altered from pink to purple upon the addition of Pb^{2+} aqueous solution, which results through the complexation mechanism of Pb^{2+} ion with Au@GSH. The developed sensor showed 1.0×10^{-1} mg L^{-1} of LOD, which the naked eye can also observe. The as-synthesized sensor was tested for the selectivity experiment with ten metal cations, proteins, and several salts. Proving the high specificity of the sensor toward Pb^{2+}, the selectivity studies showed no apparent response changes. The strong tendency of binding GSH carboxylate ions with Pb^{2+}, because of the water pH used, the protonation of $-NH_2$ groups, and the $-SH$ groups are engaged thru AuNPs avoiding interactions with Hg^{2+} ions decreasing or preventing the binding interactions significantly with ions such as Cd^{2+}, Fe^{2+}, and Zn^{2+}

ions are the key reasons of selectivity. For Pb^{2+} ion concentrations extensively, a colorimetric framework confirming the color alterations was also carried out for gradient sensitivity and quantitative analysis. In contrast to the paper-based sensors, the NFM chromic strips exhibited 28-fold better performance. A schematic representation of selectivity and sensitivity of fiber-based nanocomposites is shown in Figure 11.2.

In another study, for the sensing of carcinogenic Cr^{6+} metal ion, a hydrogel sensor coated with Fiber Bragg grating (FBG) has been synthesized. A tetraalkylammonium salt stimulus incorporated hydrogel coated on FBG is developed, precisely sensitive to Cr^{6+} ion in aqueous solutions. Thru the hydrogel's stimulus swelling property, the peak of FBG is shifted because of the mechanical strain and different concentrations of Cr^{6+} ion solutions to the developed sensor. The shift in peak is related to the Cr^{6+} metal ion concentration. Therefore, for the detection of Cr^{6+} ion, a chemo-mechanical-optical sensing method has been developed. By using this method, the Cr^{6+} ion can be detected up to 10 ppb of its trace amounts. The sensor system showed 0.83 ppb of resolution. The sensor exhibited virtuous repeatability, sensitivity, and selectivity [44].

Similarly, the SPR based optical fiber sensor has been fabricated to detect toxic metals. To synthesize the SPR sensor probe, silver (Ag) metal and ITO have been further changed with chitosan, and a pyrrole composite has been utilized.

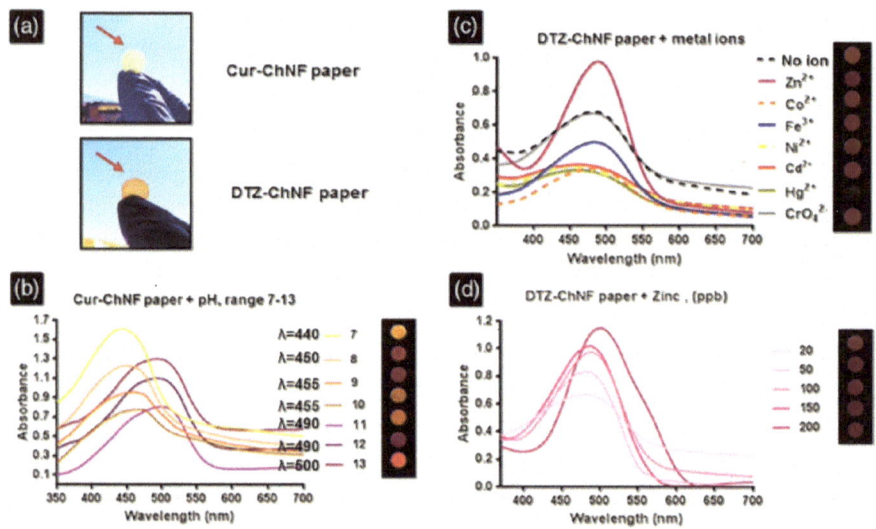

FIGURE 11.2 Diagrammatic representation of the sensing scheme using nanofiber-based optical sensors (a) ChNF-based nanocomposites (Digital photographs): Cur-ChNF paper (curcumin-ChNF paper) (top) and DTZ-ChNF paper (dithizone-ChNF paper) (bottom). (b) Cur-ChNF paper: UV–vis spectra in 7–13 pH range, (c) selectivity of DTZ-ChNF paper in a mixed metal ion solution, and (d) sensitivity upon addition of Zn^{2+} ions on DTZ-ChNF paper. (Reprinted with permission from Ref. [20]. Copyright 2020 American Chemical Society.)

They applied the wavelength examination method to the sensor to detect Hg^{2+}, Pb^{2+}, and Cd^{2+} heavy metals at their trace amounts. Among all the synthesized four sensor probes, the pyrrole/chitosan/ITO/Ag coated probe is highly sensitive. In contrast to the other ions, the Cd^{2+} ions showed greater efficiency in strongly binding with the surface of the sensing probe, which confirms the selective detection of Cd^{2+} ions. The developed sensor is best for detecting heavy metals at their low concentrations and showed that sensitivity decreases with the increasing concentration of heavy metal ions [45].

A Mach Zehnder fiber interferometer has been synthesized using a responsive hydrogel based on PVA which detects Ni^{2+} ions selectively in mixed metal ion solution. Within the hydrogel, with upsurges cross-linkages degree, the existence of Ni^{2+} increases the refractive index corresponding to the phase shift in the interferogram. The concentration of Ni^{2+} ions observed by the shifting of interference dips with 1 nM of LOD and 0.214 nm M^{-1} of sensitivity value. The capability of the gel to immobilize many receptors enables the design of the PVA hydrogel as a flexible platform for detection processes and to optimize sensitivity [46].

For the sensing of Mn^{2+} in an aqueous medium, SPR-based optical fiber sensor was developed using the coating of $(ZnO_{(1-x)}PPy_x)$ over an unclad fiber core coated with silver. The alteration in $(ZnO_{(1-x)}PPy_x)$ nanocomposites dielectric constant is used as the working principle, which occurred by Mn^{2+} in an aqueous atmosphere. Due to Mn^{2+} ions and their vital interaction with the SP wave and the sensing surface, a redshift of the SPR spectrum occurs. The shift in resonance wavelength relied on the composition ratio 'x' of $ZnO_{(1-x)}PPy_x$. The developed sensor showed high selectivity toward Mn^{2+} ions, highly toxic in drinking water because of its paramagnetic nature. When Mn^{2+} ions concentration increases, the sensitivity of the developed sensor decreases. Because of the development of probe on optical fiber, the sensor has several other applications such as remote sensing, the capability of online monitoring, and low cost [47]. Table 11.1 comprises various fiber-based optical sensors to detect heavy metal ions in aqueous media.

TABLE 11.1

The Detection of Heavy Metal Ions Using Colorimetric and Fluorescent-Based ESNFs

Analyte	Polymeric Matrix	Recognition Material	LOD (mg L^{-1})	References
Colorimetric-Based ESNFs for Heavy Metal Detection				
Hg^{2+}	PVA	AuNC	1.0×10^{-3}	[35]
	PA6/PVB	PANI	1.0×10^{-3}	[41]
Fe^{2+}	PVBC	PIMH	1.0×10^{-4} (solution) and 2.0×10^{-3} (solid state)	[25]

(Continued)

TABLE 11.1 (*Continued*)
The Detection of Heavy Metal Ions Using Colorimetric and Fluorescent-Based ESNFs

Analyte	Polymeric Matrix	Recognition Material	LOD (mg L^{-1})	References
Colorimetric-Based ESNFs for Heavy Metal Detection				
Ni^{2+}	N6	DMG	2.0×10^{-3}	[40]
Pb^{2+}	PAN	PCDA and PCDA-5EG	1.0×10^{-1}	[43]
	CA	Curcumin	4.1	[38]
	CA	PMDA	1.0×10^{-2}	[48]
	PA6/PVDF	Au@GSH NPs	1.0×10^{-1}	[43]
Fe^{3+}	Zein	Curcumin	4.0×10^{-1}	[39]
Fluorescent-Based Electrospun Nanofibers (ESNFs) for Heavy Metal Detection				
Single Metal Ion				
Fe^{3+}	PAN	CDs	0.2	[31]
	poly(HEMA-co-NMA-co-NBD)	SRhBOH	5.6	[49]
Hg^{2+}	poly(MMA-co-BNPTU-co-RhBAM)	BNPTU	4.0×10^{-3}	[50]
	PCL	AuNC	5.0×10^{-8}	[51]
	poly(NIPAAm-co-NMA-co-AA)	BNPTU	2.0×10^{-2}	[30]
	PAN	DTPDI	1.0×10^{-3}	[28]
	EC	EMIMBF4	1.4×10^{-5}	[52]
Cu^{2+}	PNNR	PNNR	6.4×10^{-3}	[53]
	poly(MMA-co-AHPA)	RhB-hydrazine	9.5×10^{-2}	[54]
	CA	DTT.AuNC	5.0×10^{-2}	[35]
	PMMA	CsPbBr3 QDs	6.0×10^{-11}	[29]
	poly(NIPAAm-co-NMA)	F-phen	-	[55]
Ni^{2+}	PAN-PAA	PAN-PAA	7.0×10^{-3}	[24]
Al^{3+}	PU	R2PP	2.0×10^{-4}	[56]
Multiple Metal Ions				
Co^{2+}/Zn^{2+}	PU	DS-5N, FL-5N and NBD-5N	2.0 for Co^{2+} and 3.0 for Zn^{2+}	[57]
Cu^{2+}/Cr^{3+}	CA	1,4-DHAQ	2.0×10^{-4} for Cu^{2+} and Cr^{2+}	[58]
Pb^{2+}/Hg^{2+}/Fe^{3+}/Mn^{2+}/Ni^{2+}/Cd^{2+}	PAN	CPEs	-	[59]
Fe^{3+}/Cr^{3+}/Hg^{2+}	PVA	SRD and SSRD	5.6×10^{-2} for Fe^{3+}, 5.2×10^{-2} for Cr^{3+} and 0.1 for Hg^{2+}	[60]

For detecting Hg^{2+} ions in an aqueous medium, a fiber optic sensor AuNPs–PVA hybrid was fabricated using AuNPs based on SPR as a sensing material. The optical properties have been studied in the films of citrate-capped AuNPs in the PVA matrix. Various analytical techniques such as FTIR, UV–vis, DLS, and TEM were used to determine the formation of AuNPs. The developed citrate-capped AuNPs in the PVA matrix showed virtuous selectivity for aqueous Hg^{2+} ions. The sensor showed $1 \times 10^{-6} M$ of detection limit toward Hg^{2+} ions [61].

For the sensing of Ni^{2+} ions selectively, an interferometric sensor with a no-core fiber (NCF) was experimentally established and functionalized with chitosan (CS)/polyacrylic acid self-assembled polyelectrolyte layers and single-mode fibers (SMF). To observe the variations in refractive index by Ni^{2+} adsorption, the sensing scheme was based on cladding modes and interference of core on the functionalized sensor. At different concentrations of Ni^{2+} ions, wavelength shifts were measured in real-time for the continuous monitoring of these metal ions. With a concentration detection limit of 0.1671 M and Ni^{2+} detection sensitivity of $0.05537 \, nm \, M^{-1}$, the developed sensor showed a linear response in the concentration range up to 500 M [62].

Pb^{2+} and Cu^{2+} ions have been detected through growing zinc oxide (ZnO) nanorods on electrospun titanium dioxide (TiO_2) nanofibers. A brush-distributed metal oxidation sensing nanostructure was developed. Persuaded through adsorbed ion concentration, a pair of inductive plane coils should measure inductive variation. The hydrothermal and electrospinning methods were used to fabricate the brush-distributed $ZnO–TiO_2$ nanofibers deposited on the PVDF-formed nanofiber membrane to improve the adsorption efficiency of Cu^{2+} and Pb^{2+} ions. The fabricated $ZnO–TiO_2$ ion sensing composites comprise the subsequent applications, such as low cost, simple fabrication process, highly effective adsorption, high selectivity, rapid response, precise porosity, surface area, and virtuous reproducibility and repeatability [63].

Similarly, for the dual-sensing of Cu^{2+} ions and dissolved oxygen (DO), a low-cost, simple method is proposed for developing a plastic optical fiber sensor. The oxygen indicator and QDs comprise optical fiber dual-sensor coated on the fiber end. For the detection of Cu^{2+} ion and DO, palladium^{2+} mesotetrakis(pentafluorophenyl)porphyrin (PdTFPP) and CdSe QDs-doped fluorinated xerogel serve as the sensing materials. Using a 405-nm LED, both the DO and Cu^{2+} ion fluorescent indicators can be excited, and the two emission wavelengths can be perceived distinctly. Regarding monitoring several DO and Cu^{2+} ion concentrations, the optical fiber dual-sensor has been experimented.

The existence of the DO sensor does not affect the luminescence properties of the Cu^{2+} ion sensor. The sensor showed a uniquely good linear response for detecting DO and Cu^{2+} ions in the 0–10 M range. In environmental, medical, and biological applications, for the detection of DO and Cu^{2+} ion concentrations simultaneously, the developed optical fiber dual-sensor can be used [64].

11.3 CONCLUSIONS AND FUTURE OUTLOOK

Electrospinning allows the formation of nanofibers with chemical functionality, altered diameter and morphology, and the opportunity of high specific surface area. The sensor capacity can be enhanced in terms of LOD and sensitivity for sensing heavy metals. Optically active ESNFs can be functionalized with carbon-based nanomaterials, conjugated polymers, and metal NPs. The physical-chemical properties of polymeric matrix and nanomaterial can be tailored into autonomously integrated and adaptable functionality to achieve high sensitivity and selectivity. The semiconducting and metallic NPs were the utmost auspicious among the many materials used for developing optically active ESNFs, because of the facile surface chemistry, wide available chemical routes for their fabrication, low cost of production, and wide-ranging electrical/optical properties. According to the target analyte, to develop sensitive and selective portable sensors, allowing naked-eye visual and fast interpretation, fluorescence and colorimetry are excellent optical transduction mechanisms when the nanofiber membranes can have their composition controlled. Some disadvantages and obstacles to be solved for extensive applications include improved control of processing parameters, an increased production rate of electrospun nanofibers, increased sensor stability, surface modification methods, and detection limits. However, electrospinning has moved from academia to industry. Despite this, fluorescent and colorimetric ESNF sensors have been discovered to be exceedingly appropriate for real sample applications toward heavy metal detection. It is supposed that in the sensor area, the advantages of such nanofibers will continue to grow in the following years.

ACKNOWLEDGMENT

Rekha Sharma gratefully acknowledges the support from the Ministry of Human Resource Development Department of Higher Education, Government of India under the scheme of Establishment of Centre of Excellence for Training and Research in FAST, for providing the necessary financial support to perform this study vide letter No, F. No. 5-5/201 4-TS. VII.

REFERENCES

1. Ying, Yulong, Wen Ying, Qiaochu Li, Donghui Meng, Guohua Ren, Rongxin Yan, and Xinsheng Peng. "Recent advances of nanomaterial-based membrane for water purification." *Applied Materials Today* 7 (2017): 144–158. doi:10.1016/j.apmt.2017.02.010.
2. March, Gregory, Tuan Dung Nguyen, and Benoit Piro. "Modified electrodes used for electrochemical detection of metal ions in environmental analysis." *Biosensors* 5, no. 2 (2015): 241–275. doi:10.3390/bios5020241.
3. Kumar, Pawan, Ki-Hyun Kim, Vasudha Bansal, Theodore Lazarides, and Naresh Kumar. "Progress in the sensing techniques for heavy metal ions using nanomaterials." *Journal of Industrial and Engineering Chemistry* 54 (2017): 30–43. doi:10.1016/j.jiec.2017.06.010.

4. Zhou, Yaoyu, Lin Tang, Guangming Zeng, Chen Zhang, Yi Zhang, and Xia Xie. "Current progress in biosensors for heavy metal ions based on DNAzymes/DNA molecules functionalized nanostructures: A review." *Sensors and Actuators B: Chemical* 223 (2016): 280–294. doi:10.1016/j.snb.2015.09.090.

5. Pandey, Shailendra Kumar, Priti Singh, Jyoti Singh, Sadhana Sachan, Sameer Srivastava, and Sunil Kumar Singh. "Nanocarbon-based electrochemical detection of heavy metals." *Electroanalysis* 28, no. 10 (2016): 2472–2488. doi:10.1002/elan.201600173.

6. Aragay, Gemma, Josefina Pons, and Arben Merkoçi. "Recent trends in macro-, micro-, and nanomaterial-based tools and strategies for heavy-metal detection." *Chemical Reviews* 111, no. 5 (2011): 3433–3458. doi:10.1021/cr100383r.

7. Mehta, Jyotsana, Sanjeev K. Bhardwaj, Neha Bhardwaj, A. K. Paul, Pawan Kumar, Ki-Hyun Kim, and Akash Deep. "Progress in the biosensing techniques for trace-level heavy metals." *Biotechnology Advances* 34, no. 1 (2016): 47–60. doi:10.1016/j.biotechadv.2015.12.001.

8. Li, Ming, Honglei Gou, Israa Al-Ogaidi, and Nianqiang Wu. "Nanostructured sensors for detection of heavy metals: A review." *ACS Sustainable Chemistry & Engineering* 1, no. 7 (2013): 713–723. doi:10.1021/sc400019a.

9. Gumpu, Manju Bhargavi, Swaminathan Sethuraman, Uma Maheswari Krishnan, and John Bosco Balaguru Rayappan. "A review on detection of heavy metal ions in water–an electrochemical approach." *Sensors and Actuators B: Chemical* 213 (2015): 515–533. doi:10.1016/j.snb.2015.02.122.

10. Lan, Lingyi, Yao Yao, Jianfeng Ping, and Yibin Ying. "Recent progress in nanomaterial-based optical aptamer assay for the detection of food chemical contaminants." *ACS Applied Materials & Interfaces* 9, no. 28 (2017): 23287–23301. doi:10.1021/acsami.7b03937.

11. Guo, Jingjing, Minjuan Zhou, and Changxi Yang. "Fluorescent hydrogel waveguide for on-site detection of heavy metal ions." *Scientific Reports* 7, no. 1 (2017): 1–8. doi:10.1038/s41598-017-08353-8.

12. Lee, Yong-Gu, Jungyoup Han, Soondong Kwon, Seoktae Kang, and Am Jang. "Development of a rotary disc voltammetric sensor system for semi-continuous and on-site measurements of Pb (II)." *Chemosphere* 143 (2016): 78–84. doi:10.1016/j.chemosphere.2015.05.069.

13. Meng, Deshuo, Nanjing Zhao, Yuanyuan Wang, Mingjun Ma, Li Fang, Yanhong Gu, Yao Jia, and Jianguo Liu. "On-line/on-site analysis of heavy metals in water and soils by laser induced breakdown spectroscopy." *Spectrochimica Acta Part B: Atomic Spectroscopy* 137 (2017): 39–45. doi:10.1016/j.sab.2017.09.011.

14. Choi, Seon-Jin, Luana Persano, Andrea Camposeo, Ji-Soo Jang, Won-Tae Koo, Sang-Joon Kim, Hee-Jin Cho, Il-Doo Kim, and Dario Pisignano. "Electrospun nanostructures for high performance chemiresistive and optical sensors." *Macromolecular Materials and Engineering* 302, no. 8 (2017): 1600569. doi:10.1002/mame.201600569.

15. Zhang, Nan, Ruirui Qiao, Jing Su, Juan Yan, Zhiqiang Xie, Yiqun Qiao, Xichang Wang, and Jian Zhong. "Recent advances of electrospun nanofibrous membranes in the development of chemosensors for heavy metal detection." *Small* 13, no. 16 (2017): 1604293. doi:10.1002/smll.201604293.

16. Schoolaert, Ella, Richard Hoogenboom, and Karen De Clerck. "Colorimetric nanofibers as optical sensors." *Advanced Functional Materials* 27, no. 38 (2017): 1702646. doi:10.1002/adfm.201702646.

17. VS, Ajay Piriya, Printo Joseph, Kiruba Daniel SCG, Susithra Lakshmanan, Takatoshi Kinoshita, and Sivakumar Muthusamy. "Colorimetric sensors for rapid detection of various analytes." *Materials Science and Engineering: C* 78 (2017): 1231–1245. doi:10.1016/j.msec.2017.05.018.

18. McDonagh, Colette, Conor S. Burke, and Brian D. MacCraith. "Optical chemical sensors." *Chemical reviews* 108, no. 2 (2008): 400–422. doi:10.1021/cr068102g.
19. Spricigo, Poliana Cristina, Milene Mitsuyuki Foschini, Caue Ribeiro, Daniel Souza Corrêa, and Marcos David Ferreira. "Nanoscaled platforms based on SiO_2 and Al_2O_3 impregnated with potassium permanganate use color changes to indicate ethylene removal." *Food and Bioprocess Technology* 10, no. 9 (2017): 1622–1630. doi:10.1007/s11947-017-1929-9.
20. Naghdi, Tina, Hamed Golmohammadi, Hossein Yousefi, Mohammad Hosseinifard, Uliana Kostiv, Daniel Horak, and Arben Merkoci. "Chitin nanofiber paper toward optical (bio) sensing applications." *ACS Applied Materials & Interfaces* 12, no. 13 (2020): 15538–15552. doi:10.1021/acsami.9b23487.
21. Roque, Aline P., Luiza A. Mercante, Vanessa P. Scagion, Juliano E. Oliveira, Luiz HC Mattoso, Leonardo De Boni, Cleber R. Mendonca, and Daniel S. Correa. "Fluorescent PMMA/MEH-PPV electrospun nanofibers: Investigation of morphology, solvent, and surfactant effect." *Journal of Polymer Science Part B: Polymer Physics* 52, no. 21 (2014): 1388–1394. doi:10.1002/polb.23574.
22. Terra, Idelma AA, Rafaela C. Sanfelice, Gustavo T. Valente, and Daniel S. Correa. "Optical sensor based on fluorescent PMMA/PFO electrospun nanofibers for monitoring volatile organic compounds." *Journal of Applied Polymer Science* 135, no. 14 (2018): 46128. doi:10.1002/app.46128.
23. Korent Urek, Špela, Nina Frančič, Matejka Turel, and Aleksandra Lobnik. "Sensing heavy metals using mesoporous-based optical chemical sensors." *Journal of Nanomaterials* 2013 (2013). doi:10.1155/2013/501320.
24. Adewuyi, Sheriff, Dezzline A. Ondigo, Ruphino Zugle, Zenixole Tshentu, Tebello Nyokong, and Nelson Torto. "A highly selective and sensitive pyridylazo-2-naphthol-poly (acrylic acid) functionalized electrospun nanofiber fluorescence "turn-off" chemosensory system for Ni^{2+}." *Analytical Methods* 4, no. 6 (2012): 1729–1735. doi:10.1039/C2AY25182E.
25. Ondigo, D. A., Z. R. Tshentu, and N. Torto. "Electrospun nanofiber based colorimetric probe for rapid detection of Fe^{2+} in water." *Analytica Chimica Acta* 804 (2013): 228–234. doi:10.1016/j.aca.2013.09.051.
26. Zhang, JingJing, FangFang Cheng, JingJing Li, Jun-Jie Zhu, and Yi Lu. "Fluorescent nanoprobes for sensing and imaging of metal ions: Recent advances and future perspectives." *Nano today* 11, no. 3 (2016): 309–329. doi:10.1016/j.nantod.2016.05.010.
27. Escudero Belmonte, Alberto, Ana Isabel Becerro Nieto, Carolina Carrillo Carrión, Nuria O. Núñez Álvarez, Mikhail V. Zyuzin, Mariano Laguna, Daniel González Mancebo, and Manuel Ocaña Jurado. "Rare earth based nanostructured materials: Synthesis, functionalization, properties and bioimaging and biosensing applications." *Nanophotonics*, 6 (5), 881–921. (2017). doi:10.1515/nanoph–2017-0007.
28. Ma, Lijing, Kelan Liu, Meizhen Yin, Jiao Chang, Yuting Geng, and Kai Pan. "Fluorescent nanofibrous membrane (FNFM) for the detection of mercuric ion (II) with high sensitivity and selectivity." *Sensors and Actuators B: Chemical* 238 (2017): 120–127. doi:10.1016/j.snb.2016.07.049.
29. Wang, Yuanwei, Yihua Zhu, Jianfei Huang, Jin Cai, Jingrun Zhu, Xiaoling Yang, Jianhua Shen, and Chunzhong Li. "Perovskite quantum dots encapsulated in electrospun fiber membranes as multifunctional supersensitive sensors for biomolecules, metal ions and pH." *Nanoscale Horizons* 2, no. 4 (2017): 225–232. doi:10.1039/C7NH00057J.
30. Liang, Fang-Cheng, Yi-Ling Luo, Chi-Ching Kuo, Bo-Yu Chen, Chia-Jung Cho, Fan-Jie Lin, Yang-Yen Yu, and Redouane Borsali. "Novel magnet and thermoresponsive chemosensory electrospinning fluorescent nanofibers and their sensing capability for metal ions." *Polymers* 9, no. 4 (2017): 136. doi:10.3390/polym9040136.

31. Li, Shouzhu, Shenghai Zhou, Hongbo Xu, Lili Xiao, Yi Wang, Hangjia Shen, Huanhuan Wang, and Qunhui Yuan. "Luminescent properties and sensing performance of a carbon quantum dot encapsulated mesoporous silica/polyacrylonitrile electrospun nanofibrous membrane." *Journal of Materials Science* 51, no. 14 (2016): 6801–6811. doi:10.1007/s10853-016-9967-7.

32. Wu, Wen-Chung, and Hsiao-Jen Lai. "Preparation of thermo-responsive electrospun nanofibers containing rhodamine-based fluorescent sensor for Cu^{2+} detection." *Journal of Polymer Research* 23, no. 11 (2016): 1–11. doi:10.1007/s10965-016-1115-1.

33. Lim, Sung H., Jonathan W. Kemling, Liang Feng, and Kenneth S. Suslick. "A colorimetric sensor array of porous pigments." *Analyst* 134, no. 12 (2009): 2453–2457. doi:10.1039/B916571A.

34. Zhang, Xin, Jun Yin, and Juyoung Yoon. "Recent advances in development of chiral fluorescent and colorimetric sensors." *Chemical Reviews* 114, no. 9 (2014): 4918–4959. doi:10.1021/cr400568b.

35. Senthamizhan, Anitha, Asli Celebioglu, Brabu Balusamy, and Tamer Uyar. "Immobilization of gold nanoclusters inside porous electrospun fibers for selective detection of Cu (II): A strategic approach to shielding pristine performance." *Scientific Reports* 5, no. 1 (2015): 1–11. doi:10.1038/srep15608.

36. Hadar, Hodayah Abuhatzira, Valery Bulatov, Bella Dolgin, and Israel Schechter. "Detection of heavy metals in water using dye nano-complexants and a polymeric film." *Journal of Hazardous Materials* 260 (2013): 652–659. doi:10.1016/j.jhazmat.2013.06.012.

37. Takahashi, Yukiko, Hitoshi Kasai, Hachiro Nakanishi, and Toshishige M. Suzuki. "Test strips for heavy-metal ions fabricated from nanosized dye compounds." *Angewandte Chemie International Edition* 45, no. 6 (2006): 913–916. doi:10.1002/anie.200503015.

38. Raj, Sarika, and Dhesingh Ravi Shankaran. "Curcumin based biocompatible nanofibers for lead ion detection." *Sensors and Actuators B: Chemical* 226 (2016): 318–325. doi:10.1016/j.snb.2015.12.006.

39. Saithongdee, Amornrat, Narong Praphairaksit, and Apichat Imyim. "Electrospun curcumin-loaded zein membrane for iron (III) ions sensing." *Sensors and Actuators B: Chemical* 202 (2014): 935–940. doi:10.1016/j.snb.2014.06.036.

40. Najarzadekan, Hamid, and Hassan Sereshti. "Development of a colorimetric sensor for nickel ion based on transparent electrospun composite nanofibers of polycaprolactam-dimethylglyoxime/polyvinyl alcohol." *Journal of Materials Science* 51, no. 18 (2016): 8645–8654. doi:10.1007/s10853-016-0123-1.

41. Si, Yang, Xueqin Wang, Yan Li, Kun Chen, Jiaqi Wang, Jianyong Yu, Hongjun Wang, and Bin Ding. "Optimized colorimetric sensor strip for mercury (II) assay using hierarchical nanostructured conjugated polymers." *Journal of Materials Chemistry A* 2, no. 3 (2014): 645–652. doi:10.1039/C3TA13867D.

42. Priyadarshini, E., and Nilotpala Pradhan. "Gold nanoparticles as efficient sensors in colorimetric detection of toxic metal ions: A review." *Sensors and Actuators B: Chemical* 238 (2017): 888–902. doi:10.1016/j.snb.2016.06.081.

43. Li, Yan, Bin Ding, Gang Sun, Tao Ke, Jingyuan Chen, Salem S. Al-Deyab, and Jianyong Yu. "Solid-phase pink-to-purple chromatic strips utilizing gold probes and nanofibrous membranes combined system for lead (II) assaying." *Sensors and Actuators B: Chemical* 204 (2014): 673–681. doi:10.1016/j.snb.2014.08.048.

44. Kishore, P. V. N., M. Sai Shankar, and M. Satyanarayana. "Detection of trace amounts of chromium (VI) using hydrogel coated Fiber Bragg grating." *Sensors and Actuators B: Chemical* 243 (2017): 626–633. doi:10.1016/j.snb.2016.12.017.

45. Verma, Roli, and Banshi D. Gupta. "Detection of heavy metal ions in contaminated water by surface plasmon resonance based optical fibre sensor using conducting polymer and chitosan." *Food Chemistry* 166 (2015): 568–575. doi:10.1016/j.foodchem.2014.06.045.

46. Tou, Z. Q., T. W. Koh, and C. C. Chan. "Poly (vinyl alcohol) hydrogel based fiber interferometer sensor for heavy metal cations." *Sensors and Actuators B: Chemical* 202 (2014): 185–193. doi:10.1016/j.snb.2014.05.006.

47. Tabassum, Rana, and Banshi D. Gupta. "Fiber optic manganese ions sensor using SPR and nanocomposite of ZnO–polypyrrole." *Sensors and Actuators B: Chemical* 220 (2015): 903–909. doi:10.1016/j.snb.2015.06.018.

48. Li, Yan, Yanan Wen, Lihuan Wang, Jianxin He, Salem S. Al-Deyab, Mohamed El-Newehy, Jianyong Yu, and Bin Ding. "Simultaneous visual detection and removal of lead (II) ions with pyromellitic dianhydride-grafted cellulose nanofibrous membranes." *Journal of Materials Chemistry A* 3, no. 35 (2015): 18180–18189. doi:10.1039/C5TA05030H.

49. Chen, Bo-Yu, Chi-Ching Kuo, Yun-Shao Huang, Shih-Tung Lu, Fang-Cheng Liang, and Dai-Hua Jiang. "Novel highly selective and reversible chemosensors based on dual-ratiometric fluorescent electrospun nanofibers with pH-and Fe^{3+}-modulated multicolor fluorescence emission." *ACS Applied Materials & Interfaces* 7, no. 4 (2015): 2797–2808. doi:10.1021/am508029x.

50. Liang, Fang-Cheng, Chi-Ching Kuo, Bo-Yu Chen, Chia-Jung Cho, Chih-Chien Hung, Wen-Chang Chen, and Redouane Borsali. "RGB-switchable porous electrospun nanofiber chemoprobe-filter prepared from multifunctional copolymers for versatile sensing of pH and heavy metals." *ACS Applied Materials & Interfaces* 9, no. 19 (2017): 16381–16396. doi:10.1021/acsami.7b00970.

51. Senthamizhan, Anitha, Asli Celebioglu, and Tamer Uyar. "Real-time selective visual monitoring of Hg^{2+} detection at ppt level: An approach to lighting electrospun nanofibers using gold nanoclusters." *Scientific Reports* 5, no. 1 (2015): 1–12. doi:10.1038/srep10403.

52. Kacmaz, Sibel, Kadriye Ertekin, Aslihan Suslu, Yavuz Ergun, Erdal Celik, and Umit Cocen. "Sub-nanomolar sensing of ionic mercury with polymeric electrospun nanofibers." *Materials Chemistry and Physics* 133, no. 1 (2012): 547–552. doi:10.1016/j.matchemphys.2012.01.081.

53. Wu, Wen-Chung, and Hsiao-Jen Lai. "Preparation of thermo-responsive electrospun nanofibers containing rhodamine-based fluorescent sensor for Cu^{2+} detection." *Journal of Polymer Research* 23, no. 11 (2016): 1–11. doi:10.1007/s10965-016-1115-1.

54. Wang, Wei, Xiuling Wang, Qingbiao Yang, Xiaoliang Fei, Mingda Sun, and Yan Song. "A reusable nanofibrous film chemosensor for highly selective and sensitive optical signaling of Cu^{2+} in aqueous media." *Chemical Communications* 49, no. 42 (2013): 4833–4835. doi:10.1039/C3CC41317A.

55. Lin, He-Jie, and Ching-Yi Chen. "Thermo-responsive electrospun nanofibers doped with 1, 10-phenanthroline-based fluorescent sensor for metal ion detection." *Journal of Materials Science* 51, no. 3 (2016): 1620–1631. doi:10.1007/s10853-015-9485-z.

56. Kim, Changkyeom, Ji-Yong Hwang, Kyo-Sun Ku, Satheshkumar Angupillai, and Young-A. Son. "A renovation of non-aqueous Al^{3+} sensor to aqueous media sensor by simple recyclable immobilize electrospun nano-fibers and its uses for live sample analysis." *Sensors and Actuators B: Chemical* 228 (2016): 259–269. doi:10.1016/j.snb.2016.01.020.

57. Anzenbacher Jr, Pavel, Fengyu Li, and Manuel A. Palacios. "Toward wearable sensors: Fluorescent attoreactor mats as optically encoded cross-reactive sensor arrays." *Angewandte Chemie* 124, no. 10 (2012): 2395–2398. doi:10.1002/ange.201105629.

58. Wang, Meiling, Guowen Meng, Qing Huang, and Yiwu Qian. "Electrospun 1, 4-DHAQ-doped cellulose nanofiber films for reusable fluorescence detection of trace Cu^{2+} and further for Cr^{3+}." *Environmental Science & Technology* 46, no. 1 (2012): 367–373. doi:10.1021/es202137c.

59. Zhang, Han, Minhua Cao, Wei Wu, Haibo Xu, Si Cheng, and Li-Juan Fan. "Polyacrylonitrile/noble metal/SiO_2 nanofibers as substrates for the amplified detection of picomolar amounts of metal ions through plasmon-enhanced fluorescence." *Nanoscale* 7, no. 4 (2015): 1374–1382. doi:10.1039/C4NR05349D.

60. Wei, Zhen, Hui Zhao, Jianhua Zhang, Liandong Deng, Siyu Wu, Junyu He, and Anjie Dong. "Poly (vinyl alcohol) electrospun nanofibrous membrane modified with spirolactam–rhodamine derivatives for visible detection and removal of metal ions." *RSC Advances* 4, no. 93 (2014): 51381–51388. doi:10.1039/C4RA07505F.

61. Raj, D. Rithesh, S. Prasanth, T. V. Vineeshkumar, and C. Sudarsanakumar. "Surface Plasmon Resonance based fiber optic sensor for mercury detection using gold nanoparticles PVA hybrid." *Optics Communications* 367 (2016): 102–107. doi:10.1016/j.optcom.2016.01.027.

62. Raghunandhan, R., L. H. Chen, H. Y. Long, L. L. Leam, P. L. So, X. Ning, and C. C. Chan. "Chitosan/PAA based fiber-optic interferometric sensor for heavy metal ions detection." *Sensors and Actuators B: Chemical* 233 (2016): 31–38. doi:10.1016/j.snb.2016.04.020.

63. Chang, Hsing-Cheng, Yu-Liang Hsu, Cheng-Yan Tsai, Ya-Hui Chen, and Shyan-Lung Lin. "Nanofiber-based brush-distributed sensor for detecting heavy metal ions." *Microsystem Technologies* 23, no. 2 (2017): 507–514. doi:10.1007/s00542-016-3231-6.

64. Chu, Cheng-Shane, and Chih-Yung Chuang. "Optical fiber sensor for dual sensing of dissolved oxygen and Cu^{2+} ions based on PdTFPP/CdSe embedded in sol–gel matrix." *Sensors and Actuators B: Chemical* 209 (2015): 94–99. doi:10.1016/j.snb.2014.11.084.

Index

Taylor & Francis eBooks

www.taylorfrancis.com

A single destination for eBooks from Taylor & Francis
with increased functionality and an improved user
experience to meet the needs of our customers.

90,000+ eBooks of award-winning academic content in
Humanities, Social Science, Science, Technology, Engineering,
and Medical written by a global network of editors and authors.

TAYLOR & FRANCIS EBOOKS OFFERS:

A streamlined
experience for
our library
customers

A single point
of discovery
for all of our
eBook content

Improved
search and
discovery of
content at both
book and
chapter level

REQUEST A FREE TRIAL
support@taylorfrancis.com

 Routledge
Taylor & Francis Group

 CRC Press
Taylor & Francis Group